James J. Buckley

Fuzzy Statistics

Springer

Berlin
Heidelberg
New York
Hong Kong
London
Milano
Paris
Tokyo

Studies in Fuzziness and Soft Computing, Volume 149

Editor-in-chief
Prof. Janusz Kacprzyk
Systems Research Institute
Polish Academy of Sciences
ul. Newelska 6
01-447 Warsaw
Poland
E-mail: kacprzyk@ibspan.waw.pl

Further volumes of this series
can be found on our homepage:
springeronline.com

Vol 130. P.S. Nair
Uncertainty in Multi-Source Databases, 2003
ISBN 3-540-03242-8

Vol 131. J.N. Mordeson, D.S. Malik, N. Kuroki
Fuzzy Semigroups, 2003
ISBN 3-540-03243-6

Vol 132. Y. Xu, D. Ruan, K. Qin, J. Liu
Lattice-Valued Logic, 2003
ISBN 3-540-40175-X

Vol. 133. Z.-Q. Liu, J. Cai, R. Buse
Handwriting Recognition, 2003
ISBN 3-540-40177-6

Vol 134. V.A. Niskanen
Soft Computing Methods in Human Sciences, 2004
ISBN 3-540-00466-1

Vol. 135. J.J. Buckley
Fuzzy Probabilities and Fuzzy Sets for Web Planning, 2004
ISBN 3-540-00473-4

Vol. 136. L. Wang (Ed.)
Soft Computing in Communications, 2004
ISBN 3-540-40575-5

Vol. 137. V. Loia, M. Nikravesh, L.A. Zadeh (Eds.)
Fuzzy Logic and the Internet, 2004
ISBN 3-540-20180-7

Vol. 138. S. Sirmakessis (Ed.)
Text Mining and its Applications, 2004
ISBN 3-540-20238-2

Vol. 139. M. Nikravesh, B. Azvine, I. Yager, L.A. Zadeh (Eds.)
Enhancing the Power of the Internet, 2004
ISBN 3-540-20237-4

Vol. 140. A. Abraham, L.C. Jain, B.J. van der Zwaag (Eds.)
Innovations in Intelligent Systems, 2004
ISBN 3-540-20265-X

Vol. 141. G.C. Onwubolu, B.V. Babu
New Optimzation Techniques in Engineering, 2004
ISBN 3-540-20167-X

Vol. 142. M. Nikravesh, L.A. Zadeh, V. Korotkikh (Eds.)
Fuzzy Partial Differential Equations and Relational Equations, 2004
ISBN 3-540-20322-2

Vol. 143. L. Rutkowski
New Soft Computing Techniques for System Modelling, Pattern Classification and Image Processing, 2004
ISBN 3-540-20584-5

Vol. 144. Z. Sun, G.R. Finnie
Intelligent Techniques in E-Commerce, 2004
ISBN 3-540-20518-7

Vol. 145. J. Gil-Aluja
Fuzzy Sets in the Management of Uncertainty, 2004
ISBN 3-540-20341-9

Vol. 146. J.A. Gámez, S. Moral, A. Salmerón (Eds.)
Advances in Bayesian Networks, 2004
ISBN 3-540-20876-3

Vol. 147. K. Watanabe, M.M.A. Hashem
New Algorithms and their Applications to Evolutionary Robots, 2004
ISBN 3-540-20901-8

Vol. 148. C. Martin-Vide, V. Mitrana, G. Păun (Eds.)
Formal Languages and Applications, 2004
ISBN 3-540-20907-7

James J. Buckley

Fuzzy Statistics

Springer

Prof. James J. Buckley
Mathematics Department
University of Alabama at Birmingham
Birmingham, AL 35294
USA
E-mail: buckley@math.uab.edu

ISBN 978-3-642-05924-7 **e-ISBN 978-3-540-39919-3**
ISSN 1434-9922

Library of Congress Cataloging-in-Publication-Data

A catalog record for this book is available from the Library of Congress.
Bibliographic information published by Die Deutsche Bibliothek.
Die Deutsche Bibliothek lists this publication in the Deutsche Nationalbibliographie; detailed bibliographic data is available in the Internet at http://dnb.ddb.de

Springer-Verlag is a part of Springer Science+Business Media
springeronline.com

Printed in Germany

Cover design: E. Kirchner, Springer-Verlag, Heidelberg
Printed on acid free paper 62/3020/M - 5 4 3 2 1 0

To Julianne and Helen.

Contents

1 Introduction 1
1.1 Introduction 1
1.2 Notation 2
1.3 Figures 3
1.4 Maple Commands 3
1.5 References 3

2 Fuzzy Sets 5
2.1 Introduction 5
2.2 Fuzzy Sets 5
2.2.1 Fuzzy Numbers 6
2.2.2 Alpha-Cuts 6
2.2.3 Inequalities 8
2.2.4 Discrete Fuzzy Sets 8
2.3 Fuzzy Arithmetic 8
2.3.1 Extension Principle 8
2.3.2 Interval Arithmetic 9
2.3.3 Fuzzy Arithmetic 10
2.4 Fuzzy Functions 11
2.4.1 Extension Principle 11
2.4.2 Alpha-Cuts and Interval Arithmetic 12
2.4.3 Differences 13
2.5 Ordering/Ranking Fuzzy Numbers 14
2.6 References 15

3 Estimate μ, Variance Known 17
3.1 Introduction 17
3.2 Fuzzy Estimation 17
3.3 Fuzzy Estimator of μ 18
3.4 References 22

4 Estimate μ, Variance Unknown **23**
4.1 Fuzzy Estimator of μ 23
4.2 References 26

5 Estimate p, Binomial Population **27**
5.1 Fuzzy Estimator of p 27
5.2 References 30

6 Estimate σ^2 from a Normal Population **31**
6.1 Introduction 31
6.2 Biased Fuzzy Estimator 31
6.3 Unbiased Fuzzy Estimator 32
6.4 References 34

7 Estimate $\mu_1 - \mu_2$, Variances Known **37**
7.1 Fuzzy Estimator 37
7.2 References 38

8 Estimate $\mu_1 - \mu_2$, Variances Unknown **39**
8.1 Introduction 39
8.2 Large Samples 39
8.3 Small Samples 39
8.3.1 Equal Variances 40
8.3.2 Unequal Variances 40
8.4 References 42

9 Estimate $d = \mu_1 - \mu_2$, Matched Pairs **43**
9.1 Fuzzy Estimator 43
9.2 References 44

10 Estimate $p_1 - p_2$, Binomial Populations **47**
10.1 Fuzzy Estimator 47
10.2 References 48

11 Estimate σ_1^2/σ_2^2, Normal Populations **49**
11.1 Introduction 49
11.2 Crisp Estimator 49
11.3 Fuzzy Estimator 50
11.4 References 51

12 Tests on μ, Variance Known **53**
12.1 Introduction 53
12.2 Non-Fuzzy Case 53
12.3 Fuzzy Case 54
12.4 One-Sided Tests 57
12.5 References 59

13 Tests on μ, Variance Unknown **61**
13.1 Introduction 61
13.2 Crisp Case 61
13.3 Fuzzy Model 62
13.3.1 $\overline{T}[\alpha]$ for Non-Positive Intervals 63
13.4 References 66

14 Tests on p for a Binomial Population **69**
14.1 Introduction 69
14.2 Non-Fuzzy Test 69
14.3 Fuzzy Test 70
14.4 References 71

15 Tests on σ^2, Normal Population **73**
15.1 Introduction 73
15.2 Crisp Hypothesis Test 73
15.3 Fuzzy Hypothesis Test 74
15.4 References 75

16 Tests μ_1 verses μ_2, Variances Known **77**
16.1 Introduction 77
16.2 Non-Fuzzy Test 77
16.3 Fuzzy Test 78
16.4 References 78

17 Test μ_1 verses μ_2, Variances Unknown **81**
17.1 Introduction 81
17.2 Large Samples 81
17.3 Small Samples 86
17.3.1 Equal Variances 86
17.3.2 Unequal Variances 87
17.4 $n_1 \neq n_2$ 88
17.5 References 90

18 Test $p_1 = p_2$, Binomial Populations **91**
18.1 Non-Fuzzy Test 91
18.2 Fuzzy Test 92
18.3 References 93

19 Test $d = \mu_1 - \mu_2$, Matched Pairs **95**
19.1 Crisp Test 95
19.2 Fuzzy Model 96
19.3 References 98

20 Test σ_1^2 verses σ_2^2, Normal Populations **99**
20.1 Crisp Test 99
20.2 Fuzzy Test 100
20.3 References 102

21 Fuzzy Correlation **103**
21.1 Introduction 103
21.2 Crisp Results 103
21.3 Fuzzy Theory 104
21.4 References 106

22 Estimation in Simple Linear Regression **107**
22.1 Introduction 107
22.2 Fuzzy Estimators 108
22.3 References 110

23 Fuzzy Prediction in Linear Regression **113**
23.1 Prediction 113
23.2 References 116

24 Hypothesis Testing in Regression **117**
24.1 Introduction 117
24.2 Tests on a 117
24.3 Tests on b 119
24.4 References 121

25 Estimation in Multiple Regression **123**
25.1 Introduction 123
25.2 Fuzzy Estimators 124
25.3 References 126

26 Fuzzy Prediction in Regression **129**
26.1 Prediction 129
26.2 References 132

27 Hypothesis Testing in Regression **133**
27.1 Introduction 133
27.2 Tests on b 133
27.3 Tests on c 135
27.4 References 137

28 Summary and Questions **139**
28.1 Summary 139
28.2 Questions 140
28.2.1 Unbiased Fuzzy Estimators 140
28.2.2 Comparing Fuzzy Numbers 140

28.2.3 No Decision Conclusion 140
28.2.4 Equal Sample Sizes 140
28.2.5 Future 141

29 Maple Commands 143
29.1 Introduction 143
29.2 Chapter 3 143
29.3 Chapter 4 144
29.4 Chapter 5 144
29.5 Chapter 6 144
29.6 Chapter 7 145
29.7 Chapter 8 145
29.8 Chapter 9 145
29.9 Chapter 10 145
29.10 Chapter 11 145
29.11 Chapter 12 146
29.12 Chapter 13 146
29.13 Chapter 14 147
29.14 Chapter 15 147
29.15 Chapter 16 148
29.16 Chapter 17 149
29.17 Chapter 18 150
29.18 Chapter 19 151
29.19 Chapter 20 151
29.20 Chapter 21 152
29.21 Chapter 22 153
29.22 Chapter 23 153
29.23 Chapter 24 154
29.24 Chapter 25 154
29.25 Chapter 26 155
29.26 Chapter 27 155

Index 157

List of Figures 163

List of Tables 167

Chapter 1

Introduction

1.1 Introduction

This book is written in four major divisions. The first part is the introductory chapters consisting of Chapters 1 and 2. In part two, Chapters 3-11, we develop fuzzy estimation. For example, in Chapter 3 we construct a fuzzy estimator for the mean of a normal distribution assuming the variance is known. More details on fuzzy estimation are in Chapter 3 and then after Chapter 3, Chapters 4-11 can be read independently. Part three, Chapters 12-20, are on fuzzy hypothesis testing. For example, in Chapter 12 we consider the test $H_0 : \mu = \mu_0$ verses $H_1 : \mu \neq \mu_0$ where μ is the mean of a normal distribution with known variance, but we use a fuzzy number (from Chapter 3) estimator of μ in the test statistic. More details on fuzzy hypothesis testing are in Chapter 12 and then after Chapter 12 Chapters 13-20 may be read independently. Part four, Chapters 21-27, are on fuzzy regression and fuzzy prediction. We start with fuzzy correlation in Chapter 21. Simple linear regression is the topic in Chapters 22-24 and Chapters 25-27 concentrate on multiple linear regression. Part two (fuzzy estimation) is used in Chapters 22 and 25; and part 3 (fuzzy hypothesis testing) is employed in Chapters 24 and 27. Fuzzy prediction is contained in Chapters 23 and 26.

A most important part of our models in fuzzy statistics is that we always start with a random sample producing crisp (non-fuzzy) data. Other authors discussing fuzzy statistics mostly begin with fuzzy data. We assume we have a random sample giving real number data $x_1, x_2, ..., x_n$ which is then used to generate our fuzzy estimators (Chapters 3 - 11). Using fuzzy estimators in hypothesis testing and regression obviously leads to fuzzy hypothesis testing and fuzzy regression.

First we need to be familiar with fuzzy sets. All you need to know about fuzzy sets for this book comprises Chapter 2. For a beginning introduction to fuzzy sets and fuzzy logic see [1]. One other item relating to fuzzy sets, needed in Chapters 12-21, 24 and 27, is also in Chapter 2: how we will determine which of the following three possibilities is true $\overline{M} < \overline{N}$, $\overline{M} > \overline{N}$ or $\overline{M} \approx \overline{N}$, for two fuzzy numbers $\overline{M}$, $\overline{N}$?

This book is based on, but expanded from, the following recent papers and publications: (1) fuzzy estimation ([2],[3]); (2) fuzzy hypothesis testing [4]; and (3) fuzzy regression and prediction [5]. The treatment of fuzzy hypothesis testing in this book is different from that in [4].

Prerequisites are a basic knowledge of crisp elementary statistics. We will cover most of elementary statistics that can be found in Chapters 6-9 in [8]. We do not discuss contingency tables, ANOVA or nonparametric statistics.

1.2 Notation

It is difficult, in a book with a lot of mathematics, to achieve a uniform notation without having to introduce many new specialized symbols. Our basic notation is presented in Chapter 2. What we have done is to have a uniform notation within each chapter. What this means is that we may use the letters "a" and "b" to represent a closed interval $[a, b]$ in one chapter but they could stand for parameters in a probability density in another chapter. We will have the following uniform notation throughout the book: (1) we place a "bar" over a letter to denote a fuzzy set ($\overline{A}$, $\overline{B}$, etc.), and all our fuzzy sets will be fuzzy subsets of the real numbers; and (2) an alpha-cut of a fuzzy set (Chapter 2) is always denoted by "α". Since we will be using α for alpha-cuts we need to change some standard notation in statistics: (1) we use β in confidence intervals; and (2) we will have γ as the significance level in hypothesis tests. So a $(1-\beta)100\%$ confidence interval means a 95% confidence interval if $\beta = 0.05$. When a confidence interval switches to being an alpha-cut of a fuzzy number (see Chapter 3), we switch from β to α. Also a hypothesis test $H_0 : \mu = 0$ verses $H_1 : \mu \neq 0$ at $\gamma = 0.05$ means given that H_0 is true, the probability of landing in the critical region is 0.05.

All fuzzy arithmetic is performed using α-cuts and interval arithmetic and not by using the extension principle (Chapter 2). Fuzzy arithmetic is needed in fuzzy hypothesis testing and fuzzy prediction.

The term "crisp" means not fuzzy. A crisp set is a regular set and a crisp number is a real number. There is a potential problem with the symbol "$\leq$". It usually means "fuzzy subset" as $\overline{A} \leq \overline{B}$ stands for $\overline{A}$ is a fuzzy subset of $\overline{B}$ (defined in Chapter 2). However, also in Chapter 2, $\overline{A} \leq \overline{B}$ means that fuzzy set $\overline{A}$ is less than or equal to fuzzy set $\overline{B}$. The meaning of the symbol "$\leq$" should be clear from its use. Also, throughout the book $\overline{x}$ will be the mean of a random sample, not a fuzzy set, and we explicitly point this out when it first arises in the book and then usually not point it out again,

Let $N(\mu, \sigma^2)$ denote the normal distribution with mean μ and variance σ^2. Critical values for the normal will be written z_γ (z_β) for hypothesis testing (confidence intervals). We have $P(X \geq z_\gamma) = \gamma$. The binomial distribution is $b(n,p)$ where n is the number of independent trials and p is the probability of a "success". Critical values for the (Student's) t distribution are t_γ (t_β) so that $P(X \geq t_\gamma) = \gamma$. Critical values for the chi-square distribution are χ^2_γ (χ^2_β) so that $P(\chi^2 \geq \chi^2_\gamma) = \gamma$. We also use $\chi^2_{L,\beta/2}$ ($\chi^2_{R,\beta/2}$) where $P(\chi^2 \leq \chi^2_{L,\beta/2}) = \beta/2$ ($P(\chi^2 \geq \chi^2_{R,\beta/2}) = \beta/2$). Critical values for the F distribution are F_γ (F_β) where $P(X \geq F_\gamma) = \gamma$. The degrees of freedom associated with the t (χ^2, F) will all be stated when they are used and will not show up as subscripts in the symbol t (χ^2, F). The reference we will be using for crisp statistics in Chapters 3 - 24 is [6]. We switch to [7] for Chapters 25 - 27.

1.3 Figures

Some of the figures, graphs of certain fuzzy numbers, in the book are difficult to obtain so they were created using different methods. Many graphs were done first in Maple [9] and then exported to $LaTeX2_\epsilon$. We did these figures first in Maple because of the "implicitplot" command in Maple. Let us explain why this command was important in this book. Suppose $\overline{X}$ is a fuzzy estimator we want to graph. Usually in this book we determine $\overline{X}$ by first calculating its α-cuts. Let $\overline{X}[\alpha] = [x_1(\alpha), x_2(\alpha)]$. So we get $x = x_1(\alpha)$ describing the left side of the triangular shaped fuzzy number $\overline{X}$ and $x = x_2(\alpha)$ describes the right side. On a graph we would have the x-axis horizontal and the y-axis vertical. α is on the y-axis between zero and one. Substituting y for α we need to graph $x = x_i(y)$, for $i = 1, 2$. But this is backwards, we usually have y a function of x. The "implicitplot" command allows us to do the correct graph with x a function of y when we have $x = x_i(y)$. All figures, except Figure 2.4, were done in Maple and then exported to $LaTeX2_\epsilon$. Figure 2.4 was constructed in $LaTeX2_\epsilon$.

1.4 Maple Commands

The Maple commands we used to create some of the figures, in Chapters 3-27, are included in Chapter 29.

1.5 References

1. J.J.Buckley and E.Eslami: An Introduction to Fuzzy Logic and Fuzzy Sets, Springer-Verlag, Heidelberg, Germany, 2002.

2. J.J.Buckley and E.Eslami: Uncertain Probabilities I: The Discrete Case, Soft Computing, 7(2003), pp. 500-505.

3. J.J.Buckley: Fuzzy Probabilities: New Approach and Applications, Physics-Verlag, Heidelberg, Germany, 2003.

4. J.J.Buckley: Fuzzy Statistics: Hypothesis Testing, Soft Computing. To appear.

5. J.J.Buckley: Fuzzy Statistics: Regression and Prediction, Soft Computing. To appear.

6. R.V.Hogg and E.A.Tanis: Probability and Statistical Inference, Sixth Edition, Prentice Hall, Upper Saddle River, N.J., 2001.

7. J.Johnston: Econometric Methods, Second Edition, McGraw-Hill, N.Y., 1972.

8. M.J.Triola: Elementary Statistics Using Excel, Second Edition, Addison-Wesley, N.Y., 2003.

9. Maple 6, Waterloo Maple Inc., Waterloo, Canada.

Chapter 2

Fuzzy Sets

2.1 Introduction

In this chapter we have collected together the basic ideas from fuzzy sets and fuzzy functions needed for the book. Any reader familiar with fuzzy sets, fuzzy numbers, the extension principle, α-cuts, interval arithmetic, and fuzzy functions may go on and have a look at Section 2.5. In Section 2.5 we discuss the method we will be using in this book to evaluate comparisons between fuzzy numbers. That is, in Section 2.5 we need to decide which one of the following three possibilities is true: $\overline{M} < \overline{N}$; $\overline{M} \approx \overline{N}$; or $\overline{M} > \overline{N}$, for fuzzy numbers $\overline{M}$ and $\overline{N}$. Section 2.5 will be employed in fuzzy hypothesis testing. A good general reference for fuzzy sets and fuzzy logic is [4] and [8].

Our notation specifying a fuzzy set is to place a "bar" over a letter. So $\overline{Y}$, $\overline{Z}$, $\overline{T}$, $\overline{CR}$, $\overline{F}$, $\overline{\chi}^2, \ldots, \overline{\mu}$, $\overline{p}$, $\overline{\sigma}^2$, $\overline{a}$, $\overline{b}$, $\overline{c}$, $\ldots$, all denote fuzzy sets.

2.2 Fuzzy Sets

If Ω is some set, then a fuzzy subset $\overline{A}$ of Ω is defined by its membership function, written $\overline{A}(x)$, which produces values in $[0,1]$ for all x in Ω. So, $\overline{A}(x)$ is a function mapping Ω into $[0,1]$. If $\overline{A}(x_0) = 1$, then we say x_0 belongs to $\overline{A}$, if $\overline{A}(x_1) = 0$ we say x_1 does not belong to $\overline{A}$, and if $\overline{A}(x_2) = 0.6$ we say the membership value of x_2 in $\overline{A}$ is 0.6. When $\overline{A}(x)$ is always equal to one or zero we obtain a crisp (non-fuzzy) subset of Ω. For all fuzzy sets $\overline{B}, \overline{C}, \ldots$ we use $\overline{B}(x), \overline{C}(x), \ldots$ for the value of their membership function at x. The fuzzy sets we will be using will be fuzzy numbers .

The term "crisp" will mean not fuzzy. A crisp set is a regular set. A crisp number is just a real number. A crisp function maps real numbers (or real vectors) into real numbers. A crisp solution to a problem is a solution involving crisp sets, crisp numbers, crisp functions, etc.

2.2.1 Fuzzy Numbers

A general definition of fuzzy number may be found in ([4],[8]), however our fuzzy numbers will be triangular (shaped) fuzzy numbers. A triangular fuzzy number $\overline{N}$ is defined by three numbers $a < b < c$ where the base of the triangle is the interval $[a, c]$ and its vertex is at $x = b$. Triangular fuzzy numbers will be written as $\overline{N} = (a/b/c)$. A triangular fuzzy number $\overline{N} = (1.2/2/2.4)$ is shown in Figure 2.1. We see that $\overline{N}(2) = 1$, $\overline{N}(1.6) = 0.5$, etc.

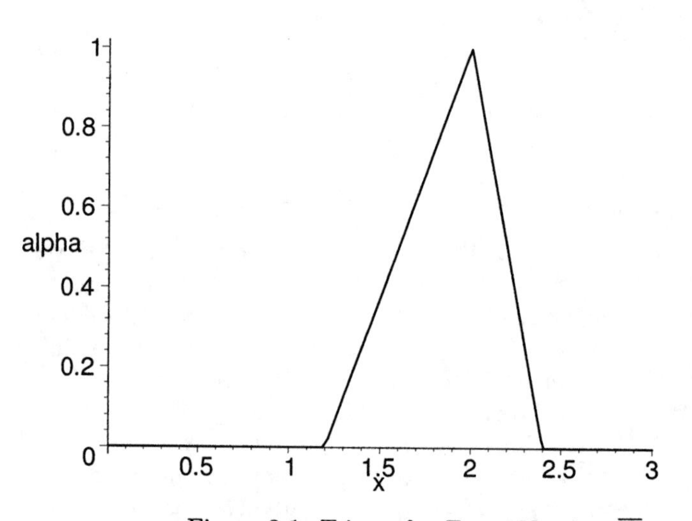

Figure 2.1: Triangular Fuzzy Number $\overline{N}$

A triangular shaped fuzzy number $\overline{P}$ is given in Figure 2.2. $\overline{P}$ is only partially specified by the three numbers 1.2, 2, 2.4 since the graph on $[1.2, 2]$, and $[2, 2.4]$, is not a straight line segment. To be a triangular shaped fuzzy number we require the graph to be continuous and: (1) monotonically increasing on $[1.2, 2]$; and (2) monotonically decreasing on $[2, 2.4]$. For triangular shaped fuzzy number $\overline{P}$ we use the notation $\overline{P} \approx (1.2/2/2.4)$ to show that it is partially defined by the three numbers 1.2, 2, and 2.4. If $\overline{P} \approx (1.2/2/2.4)$ we know its base is on the interval $[1.2, 2.4]$ with vertex (membership value one) at $x = 2$.

2.2.2 Alpha-Cuts

Alpha-cuts are slices through a fuzzy set producing regular (non-fuzzy) sets. If $\overline{A}$ is a fuzzy subset of some set Ω, then an α-cut of $\overline{A}$, written $\overline{A}[\alpha]$, is defined as

$$\overline{A}[\alpha] = \{x \in \Omega | \overline{A}(x) \geq \alpha\}, \tag{2.1}$$

for all α, $0 < \alpha \leq 1$. The $\alpha = 0$ cut, or $\overline{A}[0]$, must be defined separately.

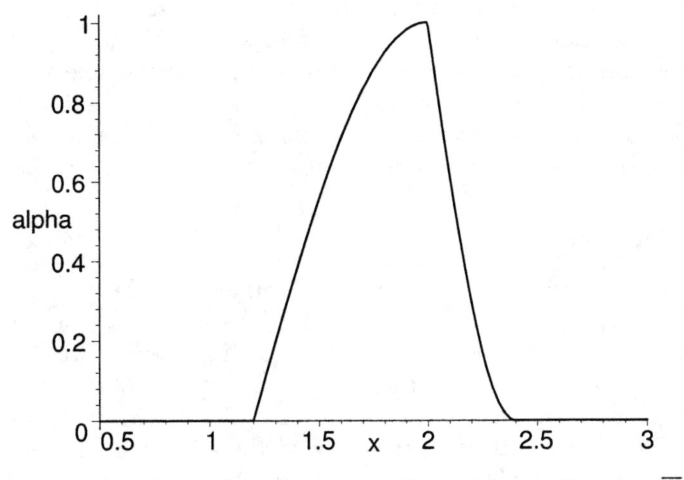

Figure 2.2: Triangular Shaped Fuzzy Number $\overline{P}$

Let $\overline{N}$ be the fuzzy number in Figure 2.1. Then $\overline{N}[0] = [1.2, 2.4]$. Notice that using equation (2.1) to define $\overline{N}[0]$ would give $\overline{N}[0]$ = all the real numbers. Similarly, in Figure 2.2 $\overline{P}[0] = [1.2, 2.4]$. For any fuzzy set $\overline{A}$, $\overline{A}[0]$ is called the support, or base, of $\overline{A}$. Many authors call the support of a fuzzy number the open interval (a, b) like the support of $\overline{N}$ in Figure 2.1 would then be $(1.2, 2.4)$. However in this book we use the closed interval $[a, b]$ for the support (base) of the fuzzy number.

The core of a fuzzy number is the set of values where the membership value equals one. If $\overline{N} = (a/b/c)$, or $\overline{N} \approx (a/b/c)$, then the core of $\overline{N}$ is the single point b.

For any fuzzy number $\overline{Q}$ we know that $\overline{Q}[\alpha]$ is a closed, bounded, interval for $0 \leq \alpha \leq 1$. We will write this as

$$\overline{Q}[\alpha] = [q_1(\alpha), q_2(\alpha)], \tag{2.2}$$

where $q_1(\alpha)$ $(q_2(\alpha))$ will be an increasing (decreasing) function of α with $q_1(1) = q_2(1)$. If $\overline{Q}$ is a triangular shaped then: (1) $q_1(\alpha)$ will be a continuous, monotonically increasing function of α in $[0, 1]$; (2) $q_2(\alpha)$ will be a continuous, monotonically decreasing function of α, $0 \leq \alpha \leq 1$; and (3) $q_1(1) = q_2(1)$.

For the $\overline{N}$ in Figure 2.1 we obtain $\overline{N}[\alpha] = [n_1(\alpha), n_2(\alpha)]$, $n_1(\alpha) = 1.2 + 0.8\alpha$ and $n_2(\alpha) = 2.4 - 0.4\alpha$, $0 \leq \alpha \leq 1$. The equation for $n_i(\alpha)$ is backwards. With the y-axis vertical and the x-axis horizontal the equation $n_1(\alpha) = 1.2 + 0.8\alpha$ means $x = 1.2 + 0.8y$, $0 \leq y \leq 1$. That is, the straight line segment from $(1.2, 0)$ to $(2, 1)$ in Figure 2.1 is given as x a function of y whereas it is usually stated as y a function of x. This is how it will be done for all α-cuts of fuzzy numbers.

The general requirements for a fuzzy set $\overline{N}$ of the real numbers to be a fuzzy number are: (1) it must be normalized, or $\overline{N}(x) = 1$ for some x; and (2) its alpha-cuts must be closed, bounded, intervals for all alpha in $[0, 1]$. This will be important in fuzzy estimation because there the fuzzy numbers will have very short vertical line segments at both ends of its base (see Section 3.2 in Chapter 3). Even so, such a fuzzy set still meets the general requirements presented above to be called a fuzzy number.

2.2.3 Inequalities

Let $\overline{N} = (a/b/c)$. We write $\overline{N} \geq \delta$, δ some real number, if $a \geq \delta$, $\overline{N} > \delta$ when $a > \delta$, $\overline{N} \leq \delta$ for $c \leq \delta$ and $\overline{N} < \delta$ if $c < \delta$. We use the same notation for triangular shaped fuzzy numbers whose support is the interval $[a, c]$.

If $\overline{A}$ and $\overline{B}$ are two fuzzy subsets of a set Ω, then $\overline{A} \leq \overline{B}$ means $\overline{A}(x) \leq \overline{B}(x)$ for all x in Ω, or $\overline{A}$ is a fuzzy subset of $\overline{B}$. $\overline{A} < \overline{B}$ holds when $\overline{A}(x) < \overline{B}(x)$, for all x. There is a potential problem with the symbol $<$. In some places in the book, for example see Section 2.5 and in hypothesis testing, $\overline{M} < \overline{N}$, for fuzzy numbers $\overline{M}$ and $\overline{N}$, means that $\overline{M}$ is less than $\overline{N}$. It should be clear on how we use "<" as to which meaning is correct.

2.2.4 Discrete Fuzzy Sets

Let $\overline{A}$ be a fuzzy subset of Ω. If $\overline{A}(x)$ is not zero only at a finite number of x values in Ω, then $\overline{A}$ is called a discrete fuzzy set. Suppose $\overline{A}(x)$ is not zero only at x_1, x_2, x_3 and x_4 in Ω. Then we write the fuzzy set as

$$\overline{A} = \{\frac{\mu_1}{x_1}, \cdots, \frac{\mu_4}{x_4}\}, \tag{2.3}$$

where the μ_i are the membership values. That is, $\overline{A}(x_i) = \mu_i$, $1 \leq i \leq 4$, and $\overline{A}(x) = 0$ otherwise. We can have discrete fuzzy subsets of any space Ω. Notice that α-cuts of discrete fuzzy sets of $\mathbb{R}$, the set of real numbers, do not produce closed, bounded, intervals.

2.3 Fuzzy Arithmetic

If $\overline{A}$ and $\overline{B}$ are two fuzzy numbers we will need to add, subtract, multiply and divide them. There are two basic methods of computing $\overline{A} + \overline{B}$, $\overline{A} - \overline{B}$, etc. which are: (1) extension principle; and (2) α-cuts and interval arithmetic.

2.3.1 Extension Principle

Let $\overline{A}$ and $\overline{B}$ be two fuzzy numbers. If $\overline{A} + \overline{B} = \overline{C}$, then the membership function for $\overline{C}$ is defined as

$$\overline{C}(z) = \sup_{x,y}\{\min(\overline{A}(x), \overline{B}(y)) | x + y = z\} . \tag{2.4}$$

If we set $\overline{C} = \overline{A} - \overline{B}$, then

$$\overline{C}(z) = \sup_{x,y}\{\min(\overline{A}(x), \overline{B}(y)) | x - y = z\} . \tag{2.5}$$

Similarly, $\overline{C} = \overline{A} \cdot \overline{B}$, then

$$\overline{C}(z) = \sup_{x,y}\{\min(\overline{A}(x), \overline{B}(y)) | x \cdot y = z\}, \tag{2.6}$$

and if $\overline{C} = \overline{A}/\overline{B}$,

$$\overline{C}(z) = \sup_{x,y}\{\min(\overline{A}(x), \overline{B}(y)) | x/y = z\} . \tag{2.7}$$

In all cases $\overline{C}$ is also a fuzzy number [8]. We assume that zero does not belong to the support of $\overline{B}$ in $\overline{C} = \overline{A}/\overline{B}$. If $\overline{A}$ and $\overline{B}$ are triangular (shaped) fuzzy numbers then so are $\overline{A}+\overline{B}$ and $\overline{A}-\overline{B}$, but $\overline{A}\cdot\overline{B}$ and $\overline{A}/\overline{B}$ will be triangular (shaped) shaped fuzzy numbers.

We should mention something about the operator "sup" in equations (2.4)-(2.7). If Ω is a set of real numbers bounded above (there is a M so that $x \leq M$, for all x in Ω), then $\sup(\Omega)$ = the least upper bound for Ω. If Ω has a maximum member, then $\sup(\Omega) = \max(\Omega)$. For example, if $\Omega = [0, 1)$, $\sup(\Omega) = 1$ but if $\Omega = [0, 1]$, then $\sup(\Omega) = \max(\Omega) = 1$. The dual operator to "sup" is "inf". If Ω is bounded below (there is a M so that $M \leq x$ for all $x \in \Omega$), then $\inf(\Omega)$ = the greatest lower bound. For example, for $\Omega = (0, 1]$ $\inf(\Omega) = 0$ but if $\Omega = [0, 1]$, then $\inf(\Omega) = \min(\Omega) = 0$.

Obviously, given $\overline{A}$ and $\overline{B}$, equations (2.4)- (2.7) appear quite complicated to compute $\overline{A}+\overline{B}$, $\overline{A}-\overline{B}$, etc. So, we now present another procedure based on α-cuts and interval arithmetic. First, we present the basics of interval arithmetic.

2.3.2 Interval Arithmetic

We only give a brief introduction to interval arithmetic. For more information the reader is referred to ([9],[10]). Let $[a_1, b_1]$ and $[a_2, b_2]$ be two closed, bounded, intervals of real numbers. If $*$ denotes addition, subtraction, multiplication, or division, then $[a_1, b_1] * [a_2, b_2] = [\alpha, \beta]$ where

$$[\alpha, \beta] = \{a * b | a_1 \leq a \leq b_1, a_2 \leq b \leq b_2\} . \tag{2.8}$$

If $*$ is division, we must assume that zero does not belong to $[a_2, b_2]$. We may simplify equation (2.8) as follows:

$$[a_1, b_1] + [a_2, b_2] = [a_1 + a_2, b_1 + b_2], \tag{2.9}$$

$$[a_1, b_1] - [a_2, b_2] = [a_1 - b_2, b_1 - a_2], \tag{2.10}$$

$$[a_1, b_1] / [a_2, b_2] = [a_1, b_1] \cdot \left[\frac{1}{b_2}, \frac{1}{a_2}\right], \tag{2.11}$$

and

$$[a_1, b_1] \cdot [a_2, b_2] = [\alpha, \beta], \tag{2.12}$$

where

$$\alpha = \min\{a_1a_2, a_1b_2, b_1a_2, b_1b_2\}, \tag{2.13}$$
$$\beta = \max\{a_1a_2, a_1b_2, b_1a_2, b_1b_2\} . \tag{2.14}$$

Multiplication and division may be further simplified if we know that $a_1 > 0$ and $b_2 < 0$, or $b_1 > 0$ and $b_2 < 0$, etc. For example, if $a_1 \geq 0$ and $a_2 \geq 0$, then

$$[a_1, b_1] \cdot [a_2, b_2] = [a_1a_2, b_1b_2], \tag{2.15}$$

and if $b_1 < 0$ but $a_2 \geq 0$, we see that

$$[a_1, b_1] \cdot [a_2, b_2] = [a_1b_2, a_2b_1] . \tag{2.16}$$

Also, assuming $b_1 < 0$ and $b_2 < 0$ we get

$$[a_1, b_1] \cdot [a_2, b_2] = [b_1b_2, a_1a_2], \tag{2.17}$$

but $a_1 \geq 0$, $b_2 < 0$ produces

$$[a_1, b_1] \cdot [a_2, b_2] = [a_2b_1, b_2a_1] . \tag{2.18}$$

2.3.3 Fuzzy Arithmetic

Again we have two fuzzy numbers $\overline{A}$ and $\overline{B}$. We know α-cuts are closed, bounded, intervals so let $\overline{A}[\alpha] = [a_1(\alpha), a_2(\alpha)]$, $\overline{B}[\alpha] = [b_1(\alpha), b_2(\alpha)]$. Then if $\overline{C} = \overline{A} + \overline{B}$ we have

$$\overline{C}[\alpha] = \overline{A}[\alpha] + \overline{B}[\alpha] . \tag{2.19}$$

We add the intervals using equation (2.9). Setting $\overline{C} = \overline{A} - \overline{B}$ we get

$$\overline{C}[\alpha] = \overline{A}[\alpha] - \overline{B}[\alpha], \tag{2.20}$$

for all α in $[0, 1]$. Also

$$\overline{C}[\alpha] = \overline{A}[\alpha] \cdot \overline{B}[\alpha], \tag{2.21}$$

for $\overline{C} = \overline{A} \cdot \overline{B}$ and

$$\overline{C}[\alpha] = \overline{A}[\alpha] / \overline{B}[\alpha], \tag{2.22}$$

when $\overline{C} = \overline{A}/\overline{B}$, provided that zero does not belong to $\overline{B}[\alpha]$ for all α. This method is equivalent to the extension principle method of fuzzy arithmetic [8]. Obviously, this procedure, of α-cuts plus interval arithmetic, is more user (and computer) friendly.

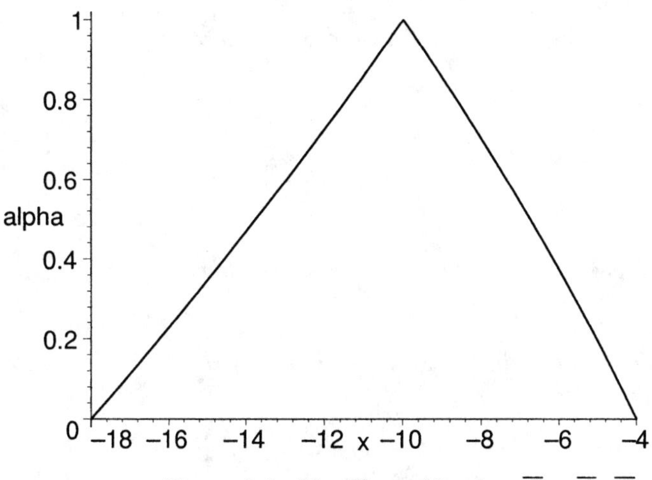

Figure 2.3: The Fuzzy Number $\overline{C} = \overline{A} \cdot \overline{B}$

Example 2.3.3.1

Let $\overline{A} = (-3/-2/-1)$ and $\overline{B} = (4/5/6)$. We determine $\overline{A} \cdot \overline{B}$ using α-cuts and interval arithmetic. We compute $\overline{A}[\alpha] = [-3+\alpha, -1-\alpha]$ and $\overline{B}[\alpha] = [4+\alpha, 6-\alpha]$. So, if $\overline{C} = \overline{A} \cdot \overline{B}$ we obtain $\overline{C}[\alpha] = [(\alpha-3)(6-\alpha), (-1-\alpha)(4+\alpha)]$, $0 \leq \alpha \leq 1$. The graph of $\overline{C}$ is shown in Figure 2.3.

2.4 Fuzzy Functions

In this book a fuzzy function is a mapping from fuzzy numbers into fuzzy numbers. We write $H(\overline{X}) = \overline{Z}$ for a fuzzy function with one independent variable $\overline{X}$. $\overline{X}$ will be a triangular (shaped) fuzzy number and then we usually obtain $\overline{Z}$ as a triangular (shaped) shaped fuzzy number. For two independent variables we have $H(\overline{X}, \overline{Y}) = \overline{Z}$.

Where do these fuzzy functions come from? They are usually extensions of real-valued functions. Let $h : [a, b] \rightarrow \mathbb{R}$. This notation means $z = h(x)$ for x in $[a, b]$ and z a real number. One extends $h : [a, b] \rightarrow \mathbb{R}$ to $H(\overline{X}) = \overline{Z}$ in two ways: (1) the extension principle; or (2) using α-cuts and interval arithmetic.

2.4.1 Extension Principle

Any $h : [a, b] \rightarrow \mathbb{R}$ may be extended to $H(\overline{X}) = \overline{Z}$ as follows

$$\overline{Z}(z) = \sup_{x} \{ \overline{X}(x) \mid h(x) = z, \; a \leq x \leq b \} \; . \tag{2.23}$$

Equation (2.23) defines the membership function of $\overline{Z}$ for any triangular (shaped) fuzzy number $\overline{X}$ in $[a, b]$.

If h is continuous, then we have a way to find α-cuts of $\overline{Z}$. Let $\overline{Z}[\alpha] = [z_1(\alpha), z_2(\alpha)]$. Then [5]

$$z_1(\alpha) = \min\{ h(x) \mid x \in \overline{X}[\alpha] \}, \tag{2.24}$$

$$z_2(\alpha) = \max\{ h(x) \mid x \in \overline{X}[\alpha] \}, \tag{2.25}$$

for $0 \leq \alpha \leq 1$.

If we have two independent variables, then let $z = h(x, y)$ for x in $[a_1, b_1]$, y in $[a_2, b_2]$. We extend h to $H(\overline{X}, \overline{Y}) = \overline{Z}$ as

$$\overline{Z}(z) = \sup_{x,y} \left\{ \min\left(\overline{X}(x), \overline{Y}(y)\right) \mid h(x, y) = z \right\}, \tag{2.26}$$

for $\overline{X}$ ($\overline{Y}$) a triangular (shaped) fuzzy number in $[a_1, b_1]$ ($[a_2, b_2]$). For α-cuts of $\overline{Z}$, assuming h is continuous, we have

$$z_1(\alpha) = \min\{ h(x, y) \mid x \in \overline{X}[\alpha],\ y \in \overline{Y}[\alpha] \}, \tag{2.27}$$

$$z_2(\alpha) = \max\{ h(x, y) \mid x \in \overline{X}[\alpha],\ y \in \overline{Y}[\alpha] \}, \tag{2.28}$$

$0 \leq \alpha \leq 1$.

2.4.2 Alpha-Cuts and Interval Arithmetic

All the functions we usually use in engineering and science have a computer algorithm which, using a finite number of additions, subtractions, multiplications and divisions, can evaluate the function to required accuracy [5]. Such functions can be extended, using α-cuts and interval arithmetic, to fuzzy functions. Let $h : [a, b] \rightarrow \mathbb{R}$ be such a function. Then its extension $H(\overline{X}) = \overline{Z}$, $\overline{X}$ in $[a, b]$ is done, via interval arithmetic, in computing $h(\overline{X}[\alpha]) = \overline{Z}[\alpha]$, α in $[0, 1]$. We input the interval $\overline{X}[\alpha]$, perform the arithmetic operations needed to evaluate h on this interval, and obtain the interval $\overline{Z}[\alpha]$. Then put these α-cuts together to obtain the value $\overline{Z}$. The extension to more independent variables is straightforward.

For example, consider the fuzzy function

$$\overline{Z} = H(\overline{X}) = \frac{\overline{A}\,\overline{X} + \overline{B}}{\overline{C}\,\overline{X} + \overline{D}}, \tag{2.29}$$

for triangular fuzzy numbers $\overline{A}$, $\overline{B}$, $\overline{C}$, $\overline{D}$ and triangular fuzzy number $\overline{X}$ in $[0, 10]$. We assume that $\overline{C} \geq 0$, $\overline{D} > 0$ so that $\overline{C}\,\overline{X} + \overline{D} > 0$. This would be the extension of

$$h(x_1, x_2, x_3, x_4, x) = \frac{x_1 x + x_2}{x_3 x + x_4}. \tag{2.30}$$

We would substitute the intervals $\overline{A}[\alpha]$ for x_1, $\overline{B}[\alpha]$ for x_2, $\overline{C}[\alpha]$ for x_3, $\overline{D}[\alpha]$ for x_4 and $\overline{X}[\alpha]$ for x, do interval arithmetic, to obtain interval $\overline{Z}[\alpha]$ for $\overline{Z}$. Alternatively, the fuzzy function

$$\overline{Z} = H(\overline{X}) = \frac{2\overline{X} + 10}{3\overline{X} + 4}, \tag{2.31}$$

would be the extension of

$$h(x) = \frac{2x + 10}{3x + 4} . \tag{2.32}$$

2.4.3 Differences

Let $h : [a, b] \to \mathbb{R}$. Just for this subsection let us write $\overline{Z}^* = H(\overline{X})$ for the extension principle method of extending h to H for $\overline{X}$ in $[a, b]$. We denote $\overline{Z} = H(\overline{X})$ for the α-cut and interval arithmetic extension of h .

We know that $\overline{Z}$ can be different from $\overline{Z}^*$. But for basic fuzzy arithmetic in Section 2.2 the two methods give the same results. In the example below we show that for $h(x) = x(1 - x)$, x in $[0, 1]$, we can get $\overline{Z}^* \neq \overline{Z}$ for some $\overline{X}$ in $[0, 1]$. What is known ([5],[9]) is that for usual functions in science and engineering $\overline{Z}^* \leq \overline{Z}$. Otherwise, there is no known necessary and sufficient conditions on h so that $\overline{Z}^* = \overline{Z}$ for all $\overline{X}$ in $[a, b]$.

There is nothing wrong in using α-cuts and interval arithmetic to evaluate fuzzy functions. Surely, it is user, and computer friendly. However, we should be aware that whenever we use α-cuts plus interval arithmetic to compute $\overline{Z} = H(\overline{X})$ we may be getting something larger than that obtained from the extension principle. The same results hold for functions of two or more independent variables.

Example 2.4.3.1

The example is the simple fuzzy expression

$$\overline{Z} = (1 - \overline{X})\,\overline{X}, \tag{2.33}$$

for $\overline{X}$ a triangular fuzzy number in $[0, 1]$. Let $\overline{X}[\alpha] = [x_1(\alpha), x_2(\alpha)]$. Using interval arithmetic we obtain

$$z_1(\alpha) = (1 - x_2(\alpha))x_1(\alpha), \tag{2.34}$$

$$z_2(\alpha) = (1 - x_1(\alpha))x_2(\alpha), \tag{2.35}$$

for $\overline{Z}[\alpha] = [z_1(\alpha), z_2(\alpha)]$, α in $[0, 1]$.

The extension principle extends the regular equation $z = (1 - x)x$, $0 \leq x \leq 1$, to fuzzy numbers as follows

$$\overline{Z}^*(z) = \sup_x \{\overline{X}(x) | (1 - x)x = z,\ 0 \leq x \leq 1\} . \tag{2.36}$$

Let $\overline{Z}^*[\alpha] = [z_1^*(\alpha), z_2^*(\alpha)]$. Then

$$z_1^*(\alpha) = \min\{(1-x)x | x \in \overline{X}[\alpha]\}, \tag{2.37}$$

$$z_2^*(\alpha) = \max\{(1-x)x | x \in \overline{X}[\alpha]\}, \tag{2.38}$$

for all $0 \le \alpha \le 1$. Now let $\overline{X} = (0/0.25/0.5)$, then $x_1(\alpha) = 0.25\alpha$ and $x_2(\alpha) = 0.50 - 0.25\alpha$. Equations (2.34) and (2.35) give $\overline{Z}[0.50] = [5/64, 21/64]$ but equations (2.37) and (2.38) produce $\overline{Z}^*[0.50] = [7/64, 15/64]$. Therefore, $\overline{Z}^* \neq \overline{Z}$. We do know that if each fuzzy number appears only once in the fuzzy expression, the two methods produce the same results ([5],[9]). However, if a fuzzy number is used more than once, as in equation (2.33), the two procedures can give different results.

2.5 Ordering/Ranking Fuzzy Numbers

Although this section is about ordering/ranking a finite set of fuzzy numbers, we will apply it to only two fuzzy numbers in fuzzy hypothesis testing.

Given a finite set of fuzzy numbers $\overline{A}_1, ..., \overline{A}_n$, we want to order/rank them from smallest to largest. For a finite set of real numbers there is no problem in ordering them from smallest to largest. However, in the fuzzy case there is no universally accepted way to do this. There are probably more than 40 methods proposed in the literature of defining $\overline{M} \le \overline{N}$, for two fuzzy numbers $\overline{M}$ and $\overline{N}$. Here the symbol $\le$ means "less than or equal" and not "a fuzzy subset of". A few key references on this topic are ([1],[6],[7],[11],[12]), where the interested reader can look up many of these methods and see their comparisons.

Here we will present only one procedure for ordering fuzzy numbers that we have used before ([2],[3]). But note that different definitions of $\le$ between fuzzy numbers can give different ordering. We first define $\le$ between two fuzzy numbers $\overline{M}$ and $\overline{N}$. Define

$$v(\overline{M} \le \overline{N}) = max\{min(\overline{M}(x), \overline{N}(y)) | x \le y\}, \tag{2.39}$$

which measures how much $\overline{M}$ is less than or equal to $\overline{N}$. We write $\overline{N} < \overline{M}$ if $v(\overline{N} \le \overline{M}) = 1$ but $v(\overline{M} \le \overline{N}) < \eta$, where η is some fixed fraction in $(0,1]$. In this book we will use $\eta = 0.8$. Then $\overline{N} < \overline{M}$ if $v(\overline{N} \le \overline{M}) = 1$ and $v(\overline{M} \le \overline{N}) < 0.8$. We then define $\overline{M} \approx \overline{N}$ when both $\overline{N} < \overline{M}$ and $\overline{M} < \overline{N}$ are false. $\overline{M} \le \overline{N}$ means $\overline{M} < \overline{N}$ or $\overline{M} \approx \overline{N}$. Now this $\approx$ may not be transitive. If $\overline{N} \approx \overline{M}$ and $\overline{M} \approx \overline{O}$ implies that $\overline{N} \approx \overline{O}$, then $\approx$ is transitive. However, it can happen that $\overline{N} \approx \overline{M}$ and $\overline{M} \approx \overline{O}$ but $\overline{N} < \overline{O}$ because $\overline{M}$ lies a little to the right of $\overline{N}$ and $\overline{O}$ lies a little to the right of $\overline{M}$ but $\overline{O}$ lies sufficiently far to the right of $\overline{N}$ that we obtain $\overline{N} < \overline{O}$. But this ordering is still useful in partitioning the set of fuzzy numbers up into sets $H_1, ..., H_K$ where ([2],[3]): (1) given any $\overline{M}$ and $\overline{N}$ in H_k, $1 \le k \le K$, then $\overline{M} \approx \overline{N}$; and (2) given $\overline{N} \in H_i$ and $\overline{M} \in H_j$, with $i < j$, then $\overline{N} \le \overline{M}$. Then the highest

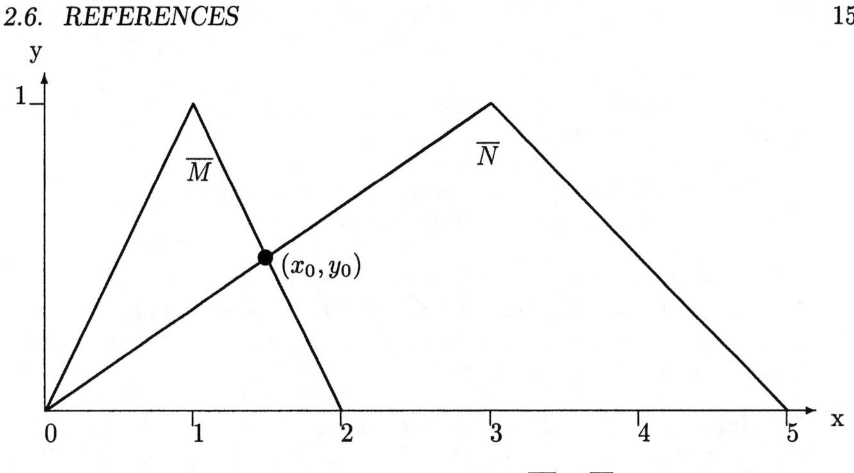

Figure 2.4: Determining $\overline{M} < \overline{N}$

ranked fuzzy numbers lie in H_K, the second highest ranked fuzzy numbers are in H_{K-1}, etc. This result is easily seen if you graph all the fuzzy numbers on the same axis then those in H_K will be clustered together farthest to the right, proceeding from the H_K cluster to the left the next cluster will be those in H_{K-1}, etc.

There is an easy way to determine if $\overline{M} < \overline{N}$, or $\overline{M} \approx \overline{N}$, for many fuzzy numbers. First, it is easy to see that if the core of $\overline{N}$ lies completely to the right of the core of $\overline{M}$, then $v(\overline{M} \leq \overline{N}) = 1$. Also, if the core of $\overline{M}$ and the core of $\overline{N}$ overlap, then $\overline{M} \approx \overline{N}$. Now assume that the core of $\overline{N}$ lies to the right of the core of $\overline{M}$, as shown in Figure 2.4 for triangular fuzzy numbers, and we wish to compute $v(\overline{N} \leq \overline{M})$. The value of this expression is simply y_0 in Figure 2.4. In general, for triangular (shaped) fuzzy numbers $v(\overline{N} \leq \overline{M})$ is the height of their intersection when the core of $\overline{N}$ lies to the right of the core of $\overline{M}$. Locate η, for example $\eta = 0.8$ in this book, on the vertical axis and then draw a horizontal line through η. If in Figure 2.4 y_0 lies below the horizontal line, then $\overline{M} < \overline{N}$. If y_0 lies on, or above, the horizontal line, then $\overline{M} \approx \overline{N}$.

2.6 References

1. G.Bortolon and R.Degani: A Review of Some Methods for Ranking Fuzzy Subsets, Fuzzy Sets and Systems, 15(1985), pp. 1-19.

2. J.J.Buckley: Ranking Alternatives Using Fuzzy Numbers, Fuzzy Sets and Systems, 15(1985), pp. 21-31.

3. J.J.Buckley: Fuzzy Hierarchical Analysis, Fuzzy Sets and Systems, 17(1985), pp. 233-247.

4. J.J.Buckley and E.Eslami: Introduction to Fuzzy Logic and Fuzzy Sets, Physica-Verlag, Heidelberg, Germany, 2002.

5. J.J.Buckley and Y.Qu: On Using α-cuts to Evaluate Fuzzy Equations, Fuzzy Sets and Systems, 38(1990), pp. 309-312.

6. P.T.Chang and E.S.Lee: Fuzzy Arithmetic and Comparison of Fuzzy Numbers, in: M.Delgado, J.Kacprzyk, J.L.Verdegay and M.A.Vila (eds.), Fuzzy Optimization: Recent Advances, Physica-Verlag, Heidelberg, Germany, 1994, pp. 69-81.

7. D.Dubois, E.Kerre, R.Mesiar and H.Prade: Fuzzy Interval Analysis, in: D.Dubois and H.Prade (eds.), Fundamentals of Fuzzy Sets, The Handbook of Fuzzy Sets, Kluwer Acad. Publ., 2000, pp. 483-581.

8. G.J.Klir and B.Yuan: Fuzzy Sets and Fuzzy Logic: Theory and Applications, Prentice Hall, Upper Saddle River, N.J., 1995.

9. R.E.Moore: Methods and Applications of Interval Analysis, SIAM Studies in Applied Mathematics, Philadelphia, 1979.

10. A.Neumaier: Interval Methods for Systems of Equations, Cambridge University Press, Cambridge, U.K., 1990.

11. X.Wang and E.E.Kerre: Reasonable Properties for the Ordering of Fuzzy Quantities (I), Fuzzy Sets and Systems, 118(2001), pp. 375-385.

12. X.Wang and E.E.Kerre: Reasonable Properties for the Ordering of Fuzzy Quantities (II), Fuzzy Sets and Systems, 118(2001), pp. 387-405.

Chapter 3

Estimate μ, Variance Known

3.1 Introduction

This starts a series of chapters, Chapters 3-11, on fuzzy estimation. In this chapter we first present some general information on fuzzy estimation and then concentrate on the mean of a normal probability distribution assuming the variance is known. The rest of the chapters on fuzzy estimation can be read independently.

3.2 Fuzzy Estimation

We will be using fuzzy numbers for estimators of parameters in probability density functions (probability mass functions in the discrete case) and in this section we show how we obtain these fuzzy numbers from a set of confidence intervals. Let X be a random variable with probability density function (or probability mass function) $f(x;\theta)$ for single parameter θ. Assume that θ is unknown and it must be estimated from a random sample $X_1, ..., X_n$. Let $Y = u(X_1, ..., X_n)$ be a statistic used to estimate θ. Given the values of these random variables $X_i = x_i$, $1 \leq i \leq n$, we obtain a point estimate $\theta^* = y = u(x_1, ..., x_n)$ for θ. We would never expect this point estimate to exactly equal θ so we often also compute a $(1-\beta)100\%$ confidence interval for θ. We are using β here since α, usually employed for confidence intervals, is reserved for α-cuts of fuzzy numbers. In this confidence interval one usually sets β equal to 0.10, 0.05 or 0.01.

We propose to find the $(1-\beta)100\%$ confidence interval for all $0.01 \leq \beta < 1$. Starting at 0.01 is arbitrary and you could begin at 0.10 or 0.05 or 0.005, etc. Denote these confidence intervals as

$$[\theta_1(\beta), \theta_2(\beta)], \tag{3.1}$$

for $0.01 \le \beta < 1$. Add to this the interval $[\theta^*, \theta^*]$ for the 0% confidence interval for θ. Then we have $(1-\beta)100\%$ confidence intervals for θ for $0.01 \le \beta \le 1$.

Now place these confidence intervals, one on top of the other, to produce a triangular shaped fuzzy number $\overline{\theta}$ whose α-cuts are the confidence intervals. We have

$$\overline{\theta}[\alpha] = [\theta_1(\alpha), \theta_2(\alpha)], \tag{3.2}$$

for $0.01 \le \alpha \le 1$. All that is needed is to finish the "bottom" of $\overline{\theta}$ to make it a complete fuzzy number. We will simply drop the graph of $\overline{\theta}$ straight down to complete its α-cuts so

$$\overline{\theta}[\alpha] = [\theta_1(0.01), \theta_2(0.01)], \tag{3.3}$$

for $0 \le \alpha < 0.01$. In this way we are using more information in $\overline{\theta}$ than just a point estimate, or just a single interval estimate.

We now have a technical problem with the membership function $\overline{\theta}(x)$ because of those vertical line segments at the two ends of its base. As it stands the graph of $\overline{\theta}$ can not be the graph of a function. Graphs of functions can not contain vertical line segments. So we will now present a special definition of the membership function $\overline{\theta}(x)$ for our fuzzy estimators. Assume that $0.01 \le \beta \le 1$. Obvious changes can be made for $0.001 \le \beta \le 1$, etc. Let $a_1 = \theta_1(0.01)$ and $a_2 = \theta_2(0.01)$ so that the base of $\overline{\theta}$ is the interval $[a_1, a_2]$. Define $\overline{\theta}(x)$ as follows: (1) $\overline{\theta}(x) = 0$ for $x < a_1$ or $x > a_2$; (2) $\overline{\theta}(x) = 0.01$ if $x = a_1$ or $x = a_2$; and (3) for $a_1 < x < a_2$ go vertically up to the graph (see Figures 3.1-3.3 below) and then horizontically over to the vertical axis to obtain the value of $\overline{\theta}(x)$. This defines the membership function but its graph will not produce the vertical line segments at the two ends of its base. We will continue to use the graph, with those two vertical line segments, as the graph of our fuzzy estimator. Extend our previous definition of a triangular shaped fuzzy number (Section 2.2.1 of Chapter 2) to cover this case and we will call $\overline{\theta}$ a triangular shaped fuzzy number. The graph of $\overline{\theta}$, containing those short vertical line segments, does satisfy the general definition of a fuzzy number given at the end of Section 2.2.2 in Chapter 2: it is normalized ($\overline{\theta}(x) = 1$ for $x = \theta^*$) and alpha-cuts are all closed, bounded, intervals. This discussion about changing the definition of the membership function for fuzzy estimators, etc. will not be repeated again in this book.

3.3 Fuzzy Estimator of μ

Consider X a random variable with probability density function $N(\mu, \sigma^2)$, which is the normal probability density with unknown mean μ and known variance σ^2. To estimate μ we obtain a random sample $X_1, ..., X_n$ from $N(\mu, \sigma^2)$. Suppose the mean of this random sample turns out to be $\overline{x}$, which is a crisp number, not a fuzzy number. We know that $\overline{x}$ is $N(\mu, \sigma^2/n)$ (Section 7.2 in [1]). So $(\overline{x} - \mu)/(\sigma/\sqrt{n})$ is $N(0,1)$. Therefore

$$P(-z_{\beta/2} \leq \frac{\overline{x} - \mu}{\sigma/\sqrt{n}} \leq z_{\beta/2}) = 1 - \beta, \tag{3.4}$$

where $z_{\beta/2}$ is the z value so that the probability of a $N(0,1)$ random variable exceeding it is $\beta/2$. Now solve the inequality for μ producing

$$P(\overline{x} - z_{\beta/2}\sigma/\sqrt{n} \leq \mu \leq \overline{x} + z_{\beta/2}\sigma/\sqrt{n}) = 1 - \beta. \tag{3.5}$$

This leads directly to the $(1-\beta)100\%$ confidence interval for μ

$$[\theta_1(\beta), \theta_2(\beta)] = [\overline{x} - z_{\beta/2}\sigma/\sqrt{n}, \overline{x} + z_{\beta/2}\sigma/\sqrt{n}], \tag{3.6}$$

where $z_{\beta/2}$ is defined as

$$\int_{-\infty}^{z_{\beta/2}} N(0,1)dx = 1 - \beta/2, \tag{3.7}$$

and $N(0,1)$ denotes the normal density with mean zero and unit variance. Put these confidence intervals together as discussed above and we obtain $\overline{\mu}$ our fuzzy estimator of μ.

The following examples show that the fuzzy estimator of the mean of the normal probability density will be a triangular shaped fuzzy number.

Example 3.3.1

Consider X a random variable with probability density function $N(\mu, 100)$, which is the normal probability density with unknown mean μ and known variance $\sigma^2 = 100$. To estimate μ we obtain a random sample $X_1, ..., X_n$ from $N(\mu, 100)$. Suppose the mean of this random sample turns out to be 28.6. Then a $(1-\beta)100\%$ confidence interval for μ is

$$[\theta_1(\beta), \theta_2(\beta)] = [28.6 - z_{\beta/2}10/\sqrt{n}, 28.6 + z_{\beta/2}10/\sqrt{n}]. \tag{3.8}$$

To obtain a graph of fuzzy μ, or $\overline{\mu}$, let $n = 64$ and first assume that $0.01 \leq \beta \leq 1$. We evaluated equations (3.8) using Maple [2] and then the final graph of $\overline{\mu}$ is shown in Figure 3.1, without dropping the graph straight down to the x-axis at the end points.

Let us go through more detail on how Maple creates the graph of the fuzzy estimator in Figure 3.1. Some information was given in Section 1.3 of Chapter 1 and the Maple commands for selected figures are in Chapter 29. In equation (3.8) the left (right) end point of the interval describes the left

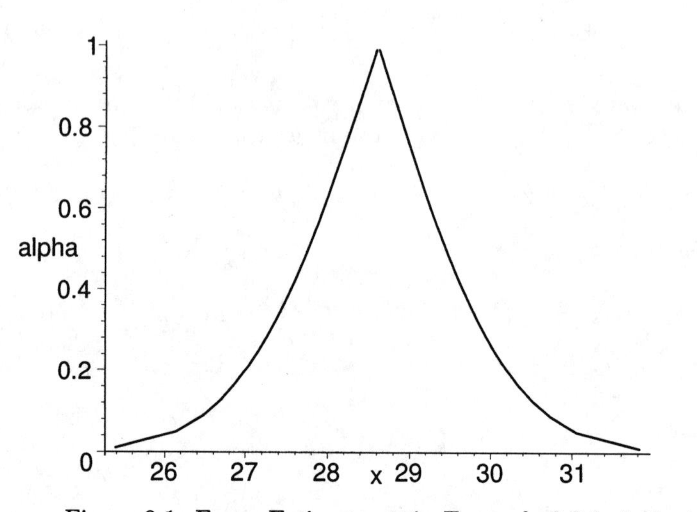

Figure 3.1: Fuzzy Estimator $\overline{\mu}$ in Example 3.3.1, $0.01 \leq \beta \leq 1$

(right) side of $\overline{\mu}$. Let the horizontal axis be called the x-axis and the vertical axis the y-axis. We now substitute y for β in equation (3.8). then

$$x = 28.6 - (1.25)z_{y/2}, \tag{3.9}$$

gives the left side of $\overline{\mu}$ and

$$x = 28.6 + (1.25)z_{y/2}, \tag{3.10}$$

is the right side of the fuzzy estimator. We have used $n = 64$ in equation (3.8). But equations (3.9) and (3.10) are "backwards" in that they give x a function of y. But using the "impliciplot" command in Maple equations (3.9) and (3.10) can be graphed and the result for $0.01 \leq y \leq 1$ is Figure 3.1.

We next evaluated equation (3.8) for $0.10 \leq \beta \leq 1$ and then the graph of $\overline{\mu}$ is shown in Figure 3.2, again without dropping the graph straight down to the x-axis at the end points. The Maple commands for Figure 3.2 are in Chapter 29. Finally, we computed equation (3.8) for $0.001 \leq \beta \leq 1$ and the graph of $\overline{\mu}$ is displayed in Figure 3.3 without dropping the graph straight down to the x-axis at the end points.

The graph in Figure 3.2 is a little misleading because the vertical axis does not start at zero. It begins at 0.08. To complete the pictures we draw short vertical line segments, from the horizontal axis up to the graph, at the end points of the base of the fuzzy number $\overline{\mu}$. The base ($\overline{\mu}[0]$) in Figure 3.1 (3.2, 3.3) is a 99% (90%, 99.9%) confidence interval for μ.

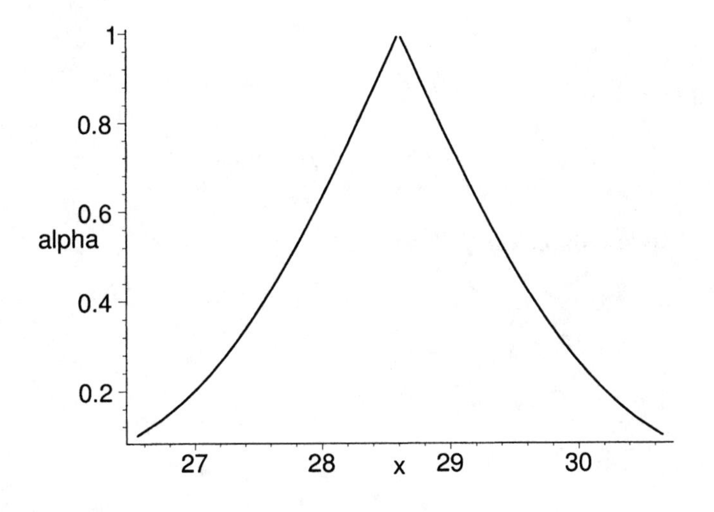

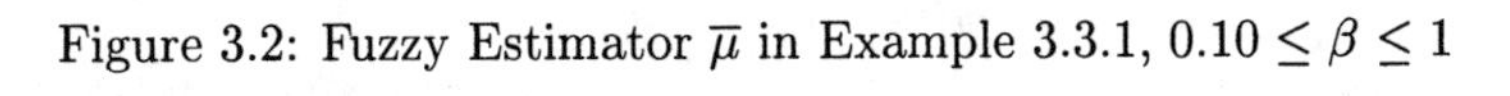

Figure 3.2: Fuzzy Estimator $\overline{\mu}$ in Example 3.3.1, $0.10 \leq \beta \leq 1$

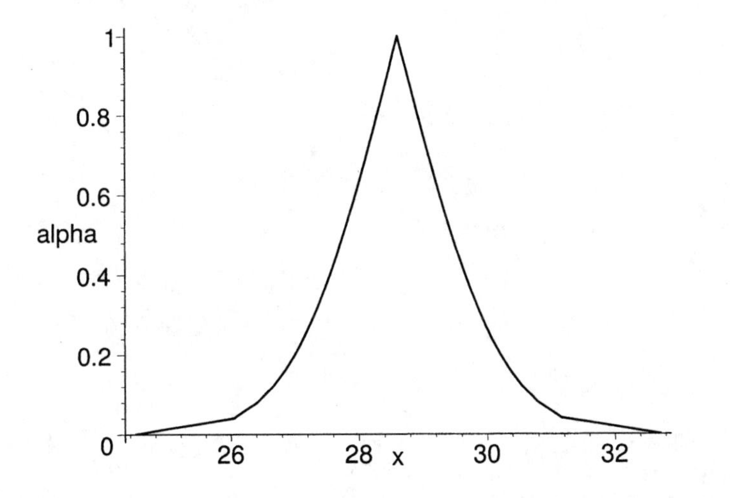

Figure 3.3: Fuzzy Estimator $\overline{\mu}$ in Example 3.3.1, $0.001 \leq \beta \leq 1$

In future chapters we usually do not explicitly mention the short vertical line segments at the two ends of the graph connecting the graph and the horizontal axis.

3.4 References

1. R.V.Hogg and E.A.Tanis: Probability and Statistical Inference, Sixth Edition, Prentice Hall, Upper Saddle River, N.J., 2001.

2. Maple 6, Waterloo Maple Inc., Waterloo, Canada.

Chapter 4

Estimate μ, Variance Unknown

4.1 Fuzzy Estimator of μ

Consider X a random variable with probability density function $N(\mu, \sigma^2)$, which is the normal probability density with unknown mean μ and unknown variance σ^2. To estimate μ we obtain a random sample $X_1, ..., X_n$ from $N(\mu, \sigma^2)$. Suppose the mean of this random sample turns out to be $\overline{x}$, which is a crisp number, not a fuzzy number. Also, let s^2 be the sample variance. Our point estimator of μ is $\overline{x}$. If the values of the random sample are $x_1, ..., x_n$ then the expression we will use for s^2 in this book is

$$s^2 = \sum_{i=1}^{n} (x_i - \overline{x})^2 / (n - 1). \tag{4.1}$$

We will use this form of s^2, with denominator $(n-1)$, so that it is an unbiased estimator of σ^2.

It is known that $(\overline{x} - \mu)/(s/\sqrt{n})$ has a (Student's) t distribution with $n - 1$ degrees of freedom (Section 7.2 of [1]). It follows that

$$P(-t_{\beta/2} \leq \frac{\overline{x} - \mu}{s/\sqrt{n}} \leq t_{\beta/2}) = 1 - \beta, \tag{4.2}$$

where $t_{\beta/2}$ is defined from the (Student's) t distribution, with $n - 1$ degrees of freedom, so that the probability of exceeding it is $\beta/2$. Now solve the inequality for μ giving

$$P(\overline{x} - t_{\beta/2} s/\sqrt{n} \leq \mu \leq \overline{x} + t_{\beta/2} s/\sqrt{n}) = 1 - \beta. \tag{4.3}$$

For this we immediately obtain the $(1 - \beta)100\%$ confidence interval for μ

$$[\overline{x} - t_{\beta/2} s/\sqrt{n}, \overline{x} + t_{\beta/2} s/\sqrt{n}]. \tag{4.4}$$

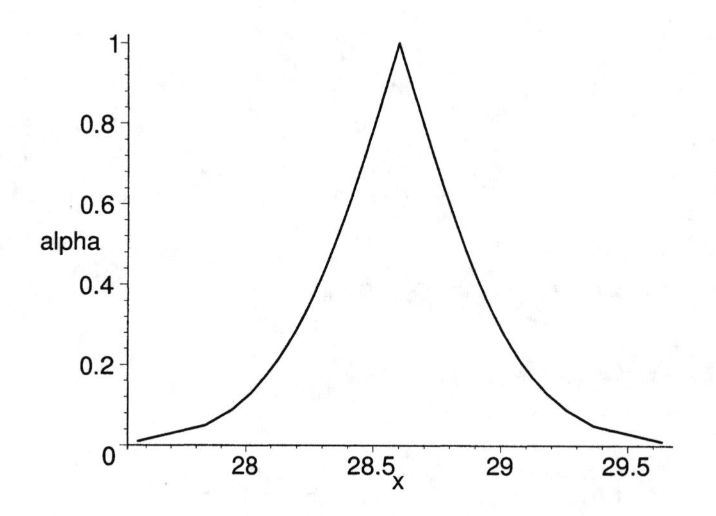

Figure 4.1: Fuzzy Estimator $\overline{\mu}$ in Example 4.1.1, $0.01 \leq \beta \leq 1$

Put these confidence intervals together, as discussed in Chapter 3, and we obtain $\overline{\mu}$ our fuzzy number estimator of μ.

Example 4.1.1

Consider X a random variable with probability density function $N(\mu, \sigma^2)$, which is the normal probability density with unknown mean μ and unknown variance σ^2. To estimate μ we obtain a random sample $X_1, ..., X_n$ from $N(\mu, \sigma^2)$. Suppose the mean of this random sample of size 25 turns out to be 28.6 and $s^2 = 3.42$. Then a $(1-\beta)100\%$ confidence interval for μ is

$$[28.6 - t_{\beta/2}\sqrt{3.42/25}, 28.6 + t_{\beta/2}\sqrt{3.42/25}]. \tag{4.5}$$

To obtain a graph of fuzzy μ, or $\overline{\mu}$, first assume that $0.01 \leq \beta \leq 1$. We evaluated equation (4.5) using Maple [2] and then the graph of $\overline{\mu}$ is shown in Figure 4.1, without dropping the graph straight down to the x-axis at the end points. The Maple commands for Figure 4.1 are in Chapter 29.

We next evaluated equation (4.5) for $0.10 \leq \beta \leq 1$ and then the graph of $\overline{\mu}$ is shown in Figure 4.2, again without dropping the graph straight down to the x-axis at the end points. Finally, we computed equation (4.5) for $0.001 \leq \beta \leq 1$ and the graph of $\overline{\mu}$ is displayed in Figure 4.3 without dropping the graph straight down to the x-axis at the end points.

The graph in Figure 4.2 is a little misleading because the vertical axis does not start at zero. It begins at 0.08. To complete the pictures we draw

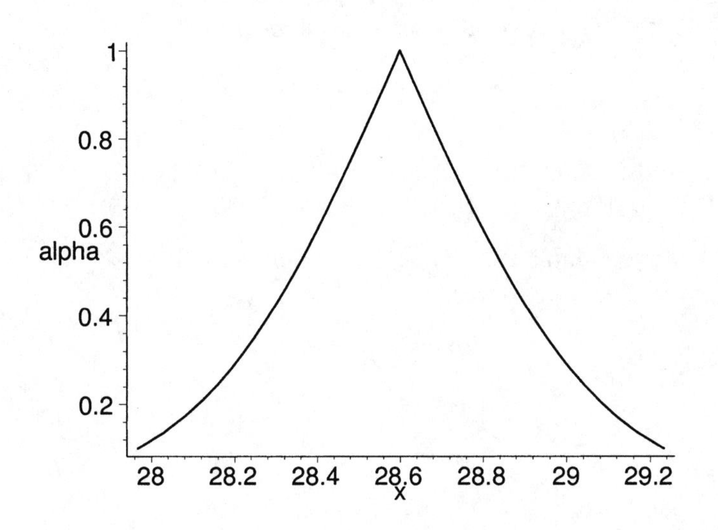

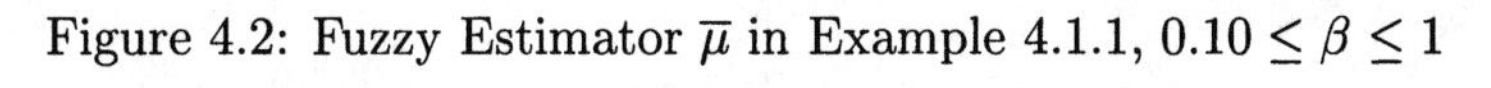
Figure 4.2: Fuzzy Estimator $\overline{\mu}$ in Example 4.1.1, $0.10 \leq \beta \leq 1$

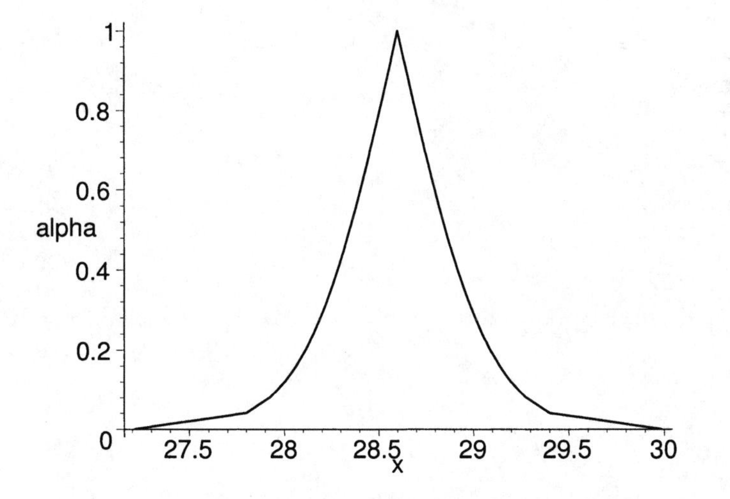

Figure 4.3: Fuzzy Estimator $\overline{\mu}$ in Example 4.1.1, $0.001 \leq \beta \leq 1$

short vertical line segments, from the horizontal axis up to the graph, at the end points of the base of the fuzzy number $\overline{\mu}$. The base ($\overline{\mu}[0]$) in Figure 4.1 (4.2, 4.3) is a 99% (90%, 99.9%) confidence interval for μ.

4.2 References

1. R.V.Hogg and E.A.Tanis: Probability and Statistical Inference, Sixth Edition, Prentice Hall, Upper Saddle River, N.J., 2001.

2. Maple 6, Waterloo Maple Inc., Waterloo, Canada.

Chapter 5

Estimate p, Binomial Population

5.1 Fuzzy Estimator of p

We have an experiment in mind in which we are interested in only two possible outcomes labeled "success" and "failure". Let p be the probability of a success so that $q = 1 - p$ will be the probability of a failure. We want to estimate the value of p. We therefore gather a random sample which here is running the experiment n independent times and counting the number of times we had a success. Let x be the number of times we observed a success in n independent repetitions of this experiment. Then our point estimate of p is $\widehat{p} = x/n$.

We know that (Section 7.5 in [1]) that $(\widehat{p} - p)/\sqrt{p(1-p)/n}$ is approximately $N(0,1)$ if n is sufficiently large. Throughout this book we will always assume that the sample size is large enough for the normal approximation to the binomial. Then

$$P(z_{\beta/2} \leq \frac{\widehat{p} - p}{\sqrt{p(1-p)/n}} \leq z_{\beta/2}) \approx 1 - \beta, \tag{5.1}$$

where $z_{\beta/2}$ was defined in equation (3.7) in Chapter 3. Solving the inequality for the p in the numerator we have

$$P(\widehat{p} - z_{\beta/2}\sqrt{p(1-p)/n} \leq p \leq \widehat{p} + z_{\beta/2}\sqrt{p(1-p)/n}) \approx 1 - \beta. \tag{5.2}$$

This leads to the $(1 - \beta)100\%$ approximate confidence interval for p

$$[\widehat{p} - z_{\beta/2}\sqrt{p(1-p)/n}, \widehat{p} + z_{\beta/2}\sqrt{p(1-p)/n}]. \tag{5.3}$$

However, we have no value for p to use in this confidence interval. So, still assuming that n is sufficiently large, we substitute $\widehat{p}$ for p in equation (5.3),

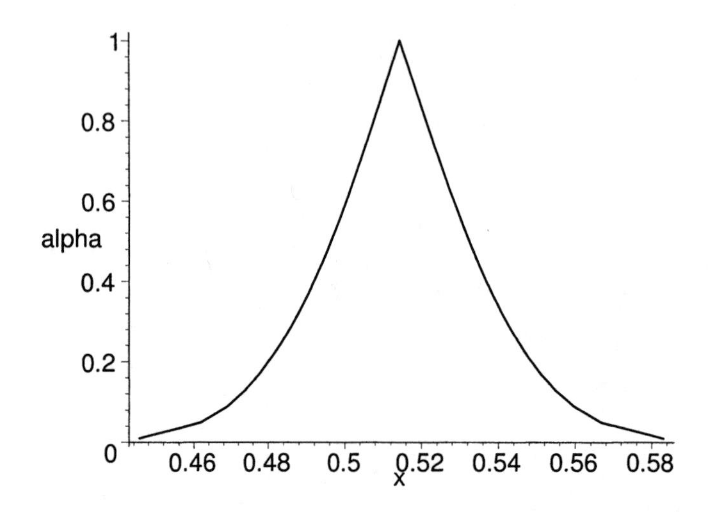

Figure 5.1: Fuzzy Estimator $\overline{p}$ in Example 5.1.1, $0.01 \leq \beta \leq 1$

using $\widehat{q} = 1 - \widehat{p}$, and we get the final $(1-\beta)100\%$ approximate confidence interval

$$[\widehat{p} - z_{\beta/2}\sqrt{\widehat{p}\widehat{q}/n}, \widehat{p} + z_{\beta/2}\sqrt{\widehat{p}\widehat{q}/n}]. \tag{5.4}$$

Put these confidence intervals together, as discussed in Chapter 3, and we get $\overline{p}$ our triangular shaped fuzzy number estimator of p.

Example 5.1.1

Assume that $n = 350$, $x = 180$ so that $\widehat{p} = 0.5143$. The confidence intervals become

$$[0.5143 - 0.0267 z_{\beta/2}, 0.5143 + 0.0267 z_{\beta/2}], \tag{5.5}$$

for $0.01(0.10, 0.001) \leq \beta \leq 1$.

To obtain a graph of fuzzy p, or $\overline{p}$, first assume that $0.01 \leq \beta \leq 1$. We evaluated equation (5.5) using Maple [2] and then the graph of $\overline{p}$ is shown in Figure 5.1, without dropping the graph straight down to the x-axis at the end points. The Maple commands for Figure 5.1 are in Chapter 29.

We next evaluated equation (5.5) for $0.10 \leq \beta \leq 1$ and then the graph of $\overline{p}$ is shown in Figure 5.2, again without dropping the graph straight down to the x-axis at the end points. Finally, we computed equation (5.5) for $0.001 \leq \beta \leq 1$ and the graph of is displayed in Figure 5.3 without dropping the graph straight down to the x-axis at the end points.

The graph in Figure 5.2 is a little misleading because the vertical axis does not start at zero. It begins at 0.08. To complete the pictures we draw

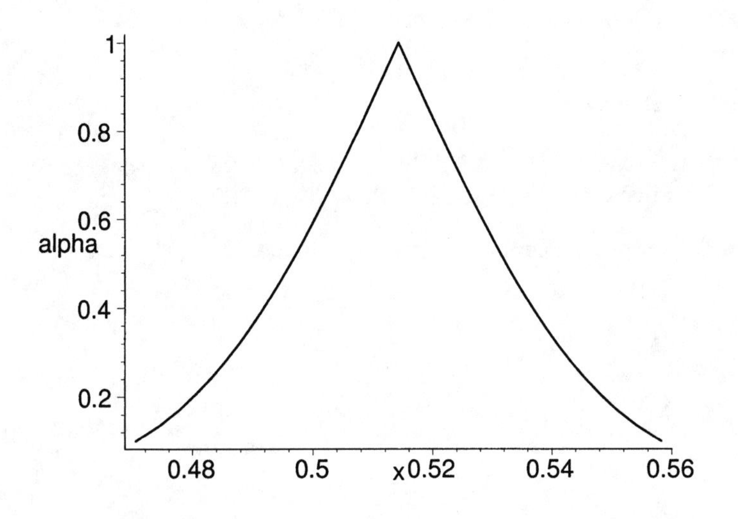

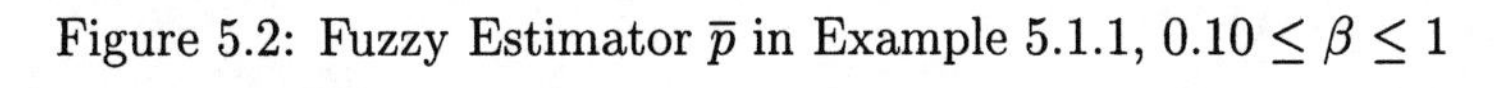
Figure 5.2: Fuzzy Estimator $\overline{p}$ in Example 5.1.1, $0.10 \leq \beta \leq 1$

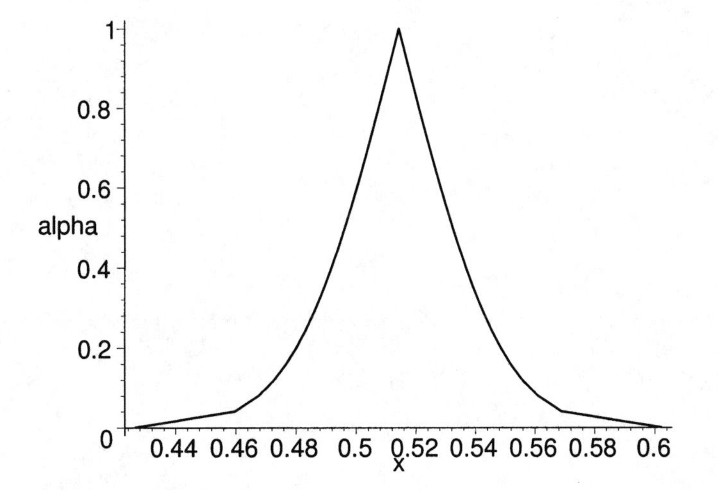

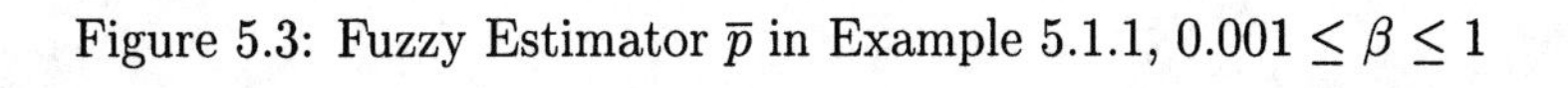
Figure 5.3: Fuzzy Estimator $\overline{p}$ in Example 5.1.1, $0.001 \leq \beta \leq 1$

short vertical line segments, from the horizontal axis up to the graph, at the end points of the base of the fuzzy number $\overline{\mu}$. The base ($\overline{\mu}[0]$) in Figure 5.1 (5.2, 5.3) is a 99% (90%, 99.9%) confidence interval for p.

5.2 References

1. R.V.Hogg and E.A.Tanis: Probability and Statistical Inference, Sixth Edition, Prentice Hall, Upper Saddle River, N.J., 2001.

2. Maple 6, Waterloo Maple Inc., Waterloo, Canada.

Chapter 6

Estimate σ^2 from a Normal Population

6.1 Introduction

We first construct a fuzzy estimator for σ^2 using the usual confidence intervals for the variance from a normal distribution and we show this fuzzy estimator is biased. Then in Section 6.3 we construct an unbiased fuzzy estimator for the variance.

6.2 Biased Fuzzy Estimator

Consider X a random variable with probability density function $N(\mu, \sigma^2)$, which is the normal probability density with unknown mean μ and unknown variance σ^2. To estimate σ^2 we obtain a random sample $X_1, ..., X_n$ from $N(\mu, \sigma^2)$. Our point estimator for the variance will be s^2. If the values of the random sample are $x_1, ..., x_n$ then the expression we will use for s^2 in this book is

$$s^2 = \sum_{i=1}^{n} (x_i - \overline{x})^2 / (n-1). \tag{6.1}$$

We will use this form of s^2, with denominator $(n-1)$, so that it is an unbiased estimator of σ^2.

We know that (Section 7.4 in [1]) $(n-1)s^2/\sigma^2$ has a chi-square distribution with $n-1$ degrees of freedom. Then

$$P(\chi^2_{L,\beta/2} \leq (n-1)s^2/\sigma^2 \leq \chi^2_{R,\beta/2}) = 1 - \beta, \tag{6.2}$$

where $\chi^2_{R,\beta/2}$ ($\chi^2_{L,\beta/2}$) is the point on the right (left) side of the χ^2 density where the probability of exceeding (being less than) it is $\beta/2$. The χ^2 distri-

n	$factor$
10	1.0788
20	1.0361
50	1.0138
100	1.0068
500	1.0013
1000	1.0007

Table 6.1: Values of $factor$ for Various Values of n

bution has $n-1$ degrees of freedom. Solve the inequality for σ^2 and we see that

$$P(\frac{(n-1)s^2}{\chi^2_{R,\beta/2}} \leq \sigma^2 \leq \frac{(n-1)s^2}{\chi^2_{L,\beta/2}}) = 1-\beta. \tag{6.3}$$

From this we obtain the usual $(1-\beta)100\%$ confidence intervals for σ^2

$$[(n-1)s^2/\chi^2_{R,\beta/2}, (n-1)s^2/\chi^2_{L,\beta/2}]. \tag{6.4}$$

Put these confidence intervals together, as discussed in Chapter 3, and we obtain $\overline{\sigma}^2$ our fuzzy number estimator of σ^2.

We now show that this fuzzy estimator is biased because the vertex of the triangular shaped fuzzy number $\overline{\sigma}^2$, where the membership value equals one, is not at s^2. We say a fuzzy estimator is biased when its vertex is not at the point estimator. We obtain the vertex of $\overline{\sigma}^2$ when $\beta = 1.0$. Let

$$factor = \frac{n-1}{\chi^2_{R,0.50}} = \frac{n-1}{\chi^2_{L,0.50}}, \tag{6.5}$$

after we substitute $\beta = 1$. Then the 0% confidence interval for the variance is

$$[(factor)(s^2), (factor)(s^2)] = (factor)(s^2). \tag{6.6}$$

Since $factor \neq 1$ the fuzzy number $\overline{\sigma}^2$ is not centered at s^2. Table 1 shows some values of $factor$ for various choices for n. We see that $factor \to 1$ as $n \to \infty$ but $factor$ is substantially larger than one for small values on n. This fuzzy estimator is biased and we will construct an unbiased (vertex at s^2) in the next section.

6.3 Unbiased Fuzzy Estimator

In deriving the usual confidence interval for the variance we start with recognizing that $(n-1)s^2/\sigma^2$ has a χ^2 distribution with $n-1$ degrees of freedom. Then for a $(1-\beta)100\%$ confidence interval we may find a and b so that

$$P(a \leq \frac{(n-1)s^2}{\sigma^2} \leq b) = 1-\beta. \tag{6.7}$$

The usual confidence interval has a and b so that the probabilities in the "two tails" are equal. That is, $a = \chi^2_{L,\beta/2}$ ($b = \chi^2_{R,\beta/2}$) so that the probability of being less (greater) than a (b) is $\beta/2$. But we do not have to pick the a and b this way ([1], p. 378). We will change the way we pick the a and b so that the fuzzy estimator is unbiased.

Assume that $0.01 \leq \beta \leq 1$. Now this interval for β is fixed and also n and s^2 are fixed. Define

$$L(\lambda) = [1-\lambda]\chi^2_{R,0.005} + \lambda(n-1), \tag{6.8}$$

and

$$R(\lambda) = [1-\lambda]\chi^2_{L,0.005} + \lambda(n-1). \tag{6.9}$$

Then a confidence interval for the variance is

$$[\frac{(n-1)s^2}{L(\lambda)}, \frac{(n-1)s^2}{R(\lambda)}], \tag{6.10}$$

for $0 \leq \lambda \leq 1$. We start with a 99% confidence interval when $\lambda = 0$ and end up with a 0% confidence interval for $\lambda = 1$. Notice that now the 0% confidence interval is $[s^2, s^2] = s^2$ and it is unbiased. As usual, we place these confidence intervals one on top of another to obtain our (unbiased) fuzzy estimator $\overline{\sigma}^2$ for the variance. Our confidence interval for σ, the population standard deviation, is

$$[\sqrt{(n-1)/L(\lambda)}s, \sqrt{(n-1)/R(\lambda)}s]. \tag{6.11}$$

Let us compare the methods in this section to those in Section 6.2. Let χ^2 be the chi-square probability density with $n-1$ degrees of freedom. The mean of χ^2 is $n-1$ and the median is the point md where $P(X \leq md) = P(X \geq md) = 0.5$. We assume β is in the interval $[0.01, 1]$. In Section 6.2 as β continuously increases from 0.01 to 1, $\chi^2_{L,\beta/2}$ ($\chi^2_{R,\beta/2}$) starts at $\chi^2_{L,0.005}$ ($\chi^2_{R,0.005}$) and increases (decreases) to $\chi^2_{L,0.5}$ ($\chi^2_{R,0.5}$) which equals to the median. Recall that $\chi^2_{L,\beta/2}$ ($\chi^2_{R,\beta/2}$) is the point on the χ^2 density where the probability of being less (greater) that it equals $\beta/2$. From Table 6.1 we see that the median is always less than $n-1$. This produces the bias in the fuzzy estimator in that section. In this section as λ continuously increases from zero to one $L(\lambda)$ ($R(\lambda)$) decreases (increases) from $\chi^2_{R,0.005}$ ($\chi^2_{L,0.005}$) to $n-1$. At $\lambda = 1$ we get $L(1) = R(1) = n-1$ and the vertex (membership value one) is at s^2 and it is now unbiased.

We will use this fuzzy estimator $\overline{\sigma}^2$ constructed in this section for σ^2 in the rest of this book. Given a value of $\lambda = \lambda^* \in [0,1]$ one may wonder what is the corresponding value of β for the confidence interval. We now show how to get the β. Let $L^* = L(\lambda^*)$ and $R^* = R(\lambda^*)$. Define

$$l = \int_0^{R^*} \chi^2 dx, \tag{6.12}$$

and

$$r = \int_{L^*}^{\infty} \chi^2 dx, \tag{6.13}$$

and then $\beta = l + r$. Now l (r) need not equal $\beta/2$. Both of these integrals above are easily evaluated using Maple [2]. The chi-square density inside these integrals has $n-1$ degrees of freedom.

Example 6.3.1

Consider X a random variable with probability density function $N(\mu, \sigma^2)$, which is the normal probability density with mean μ and unknown variance σ^2. To estimate σ^2 we obtain a random sample $X_1, ..., X_n$ from $N(\mu, \sigma^2)$. Suppose $n = 25$ and we calculate $s^2 = 3.42$. Then a confidence interval for σ^2 is

$$[\frac{82.08}{L(\lambda)}, \frac{82.08}{R(\lambda)}]. \tag{6.14}$$

To obtain a graph of fuzzy σ^2, or $\overline{\sigma}^2$, first assume that $0.01 \leq \beta \leq 1$. We evaluated equation (6.14) using Maple [2] and then the graph of $\overline{\sigma}^2$ is shown in Figure 6.1, without dropping the graph straight down to the x-axis at the end points. The Maple commands for Figure 6.1 are in Chapter 29.

We next evaluated equation (6.14) for $0.10 \leq \beta \leq 1$ and then the graph of $\overline{\sigma}^2$ is shown in Figure 6.2, again without dropping the graph straight down to the x-axis at the end points. Finally, we computed equation (6.14) for $0.001 \leq \beta \leq 1$ and the graph of $\overline{\sigma}^2$ is displayed in Figure 6.3 without dropping the graph straight down to the x-axis at the end points.

The graph in Figure 6.2 is a little misleading because the vertical axis does not start at zero. It begins at 0.08. To complete the pictures we draw short vertical line segments, from the horizontal axis up to the graph, at the end points of the base of the fuzzy number $\overline{\sigma}^2$. The base ($\overline{\sigma}^2[0]$) in Figure 6.1 (6.2, 6.3) is a 99% (90%, 99.9%) confidence interval for σ^2.

To complete this chapter let us present one graph of our fuzzy estimator $\overline{\sigma}$ of σ. Alpha-cuts of $\overline{\sigma}$ are

$$[\frac{9.06}{\sqrt{L(\lambda)}}, \frac{9.06}{\sqrt{R(\lambda)}}]. \tag{6.15}$$

Assuming $0.01 \leq \beta \leq 1$, the graph is in Figure 6.4.

6.4 References

1. R.V.Hogg and E.A.Tanis: Probability and Statistical Inference, Sixth Edition, Prentice Hall, Upper Saddle River, N.J., 2001.

2. Maple 6, Waterloo Maple Inc., Waterloo, Canada.

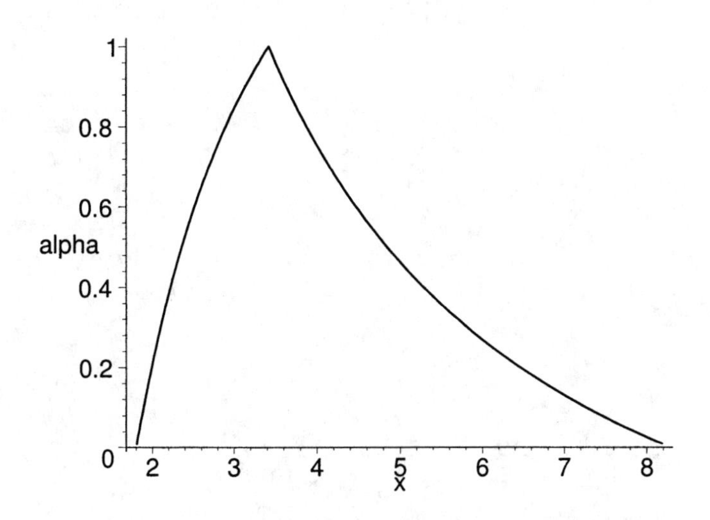

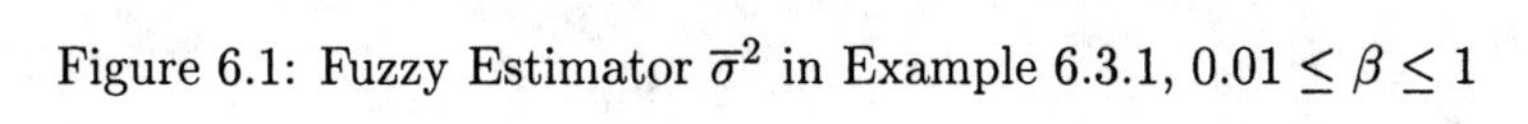
Figure 6.1: Fuzzy Estimator $\overline{\sigma}^2$ in Example 6.3.1, $0.01 \leq \beta \leq 1$

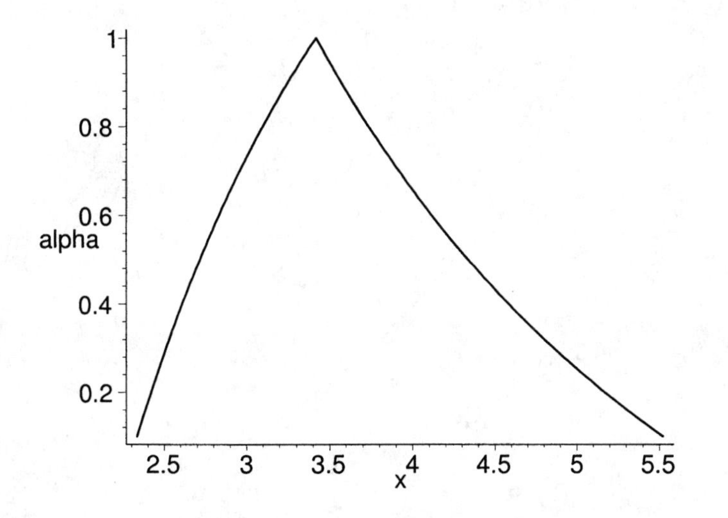

Figure 6.2: Fuzzy Estimator $\overline{\sigma}^2$ in Example 6.3.1, $0.10 \leq \beta \leq 1$

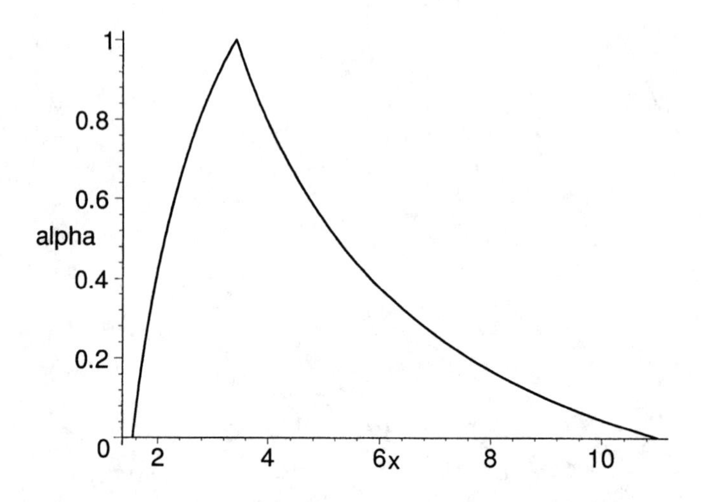

Figure 6.3: Fuzzy Estimator $\overline{\sigma}^2$ in Example 6.3.1, $0.001 \leq \beta \leq 1$

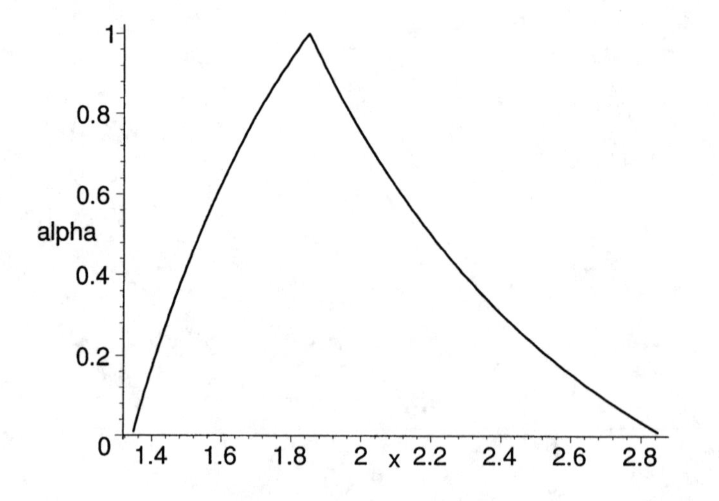

Figure 6.4: Fuzzy Estimator $\overline{\sigma}$ in Example 6.3.1, $0.01 \leq \beta \leq 1$

Chapter 7

Estimate $\mu_1 - \mu_2$, Variances Known

7.1 Fuzzy Estimator

We have two populations: Pop I and Pop II. Pop I is normally distributed with unknown mean μ_1 and known variance σ_1^2. Pop II is also normally distributed with unknown mean μ_2 but known variance σ_2^2. We wish to construct a fuzzy estimator for $\mu_1 - \mu_2$.

We collect a random sample of size n_1 from Pop I and let $\overline{x}_1$ be the mean for this data. We also gather a random sample of size n_2 from Pop II and $\overline{x}_2$ is the mean for the second sample. We assume these two random samples are independent.

Now $\overline{x}_1 - \overline{x}_2$ is normally distributed with mean $\mu_1 - \mu_2$ and standard deviation $\sigma_0 = \sqrt{\sigma_1^2/n_1 + \sigma_2^2/n_2}$ (Section 7.3 of [1]). Then as in Section 3.3 of Chapter 3, a $(1-\beta)100\%$ confidence interval for $\mu_1 - \mu_2$, is

$$[\overline{x}_1 - \overline{x}_2 - z_{\beta/2}\sigma_0, \overline{x}_1 - \overline{x}_2 + z_{\beta/2}\sigma_0], \tag{7.1}$$

where $z_{\beta/2}$ was defined in equation (3.7) in Chapter 3. So, we place these confidence intervals one on top of another to build our fuzzy estimator $\overline{\mu}_{12}$ for $\mu_1 - \mu_2$.

Example 7.1.1

Assume that: (1) $n_1 = 15$, $\overline{x}_1 = 70.1$,$\sigma_1^2 = 6$; and (2) $n_2 = 8$, $\overline{x}_2 = 75.3$,$\sigma_2^2 = 4$. Then equation (7.1) becomes

$$[-5.2 - 0.9487 z_{\beta/2}, -5.2 + 0.9487 z_{\beta/2}]. \tag{7.2}$$

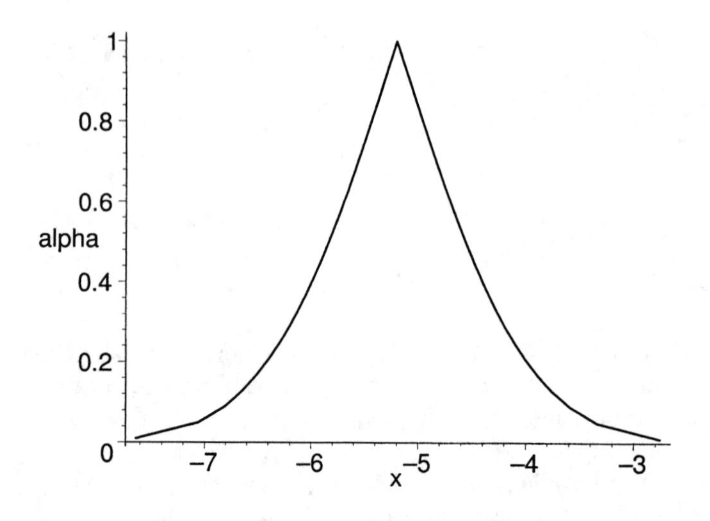

Figure 7.1: Fuzzy Estimator $\overline{\mu}_{12}$ in Example 7.1.1, $0.01 \leq \beta \leq 1$

Using Maple [2] the graph of $\overline{\mu}_{12}$ is shown in Figure 7.1 for $0.01 \leq \beta \leq 1$. We will omit other graphs of $\overline{\mu}_{12}$ for $0.10 \leq \beta \leq 1$ and $0.001 \leq \beta \leq 1$. In fact, we will only work with $0.01 \leq \beta \leq 1$, unless explicitly stated otherwise, in the rest of the book.

7.2 References

1. R.V.Hogg and E.A.Tanis: Probability and Statistical Inference, Sixth Edition, Prentice Hall, Upper Saddle River, N.J., 2001.

2. Maple 6, Waterloo Maple Inc., Waterloo, Canada.

Chapter 8

Estimate $\mu_1 - \mu_2$, Variances Unknown

8.1 Introduction

This continues Chapter 7 but now the population variances are unknown. We will use the notation of Chapter 7. There are three cases to look at and we will discuss each of these below.

8.2 Large Samples

We assume that $n_1 > 30$ and $n_2 > 30$. Let s_1^2 (s_2^2) be the sample variance calculated from the data acquired from Pop I (Pop II). With large samples we may use the normal approximation and a $(1-\beta)100\%$ confidence interval for $\mu_1 - \mu_2$ is (Section 7.3 in [1])

$$[\overline{x}_1 - \overline{x}_2 - z_{\beta/2}s_0, \overline{x}_1 - \overline{x}_2 + z_{\beta/2}s_0], \tag{8.1}$$

where $s_0 = \sqrt{s_1^2/n_1 + s_2^2/n_2}$. Put these confidence intervals together to obtain a fuzzy estimator $\overline{\mu}_{12}$ for the difference of the means. The results are similar to those in Chapter 7 so we will not present any graphs of this fuzzy estimator.

8.3 Small Samples

Here we have two cases: (1) if we may assume that the variances are equal; or (2) the variances are not equal. We assume that $n_1 \leq 30$ and/or $n_2 \leq 30$.

8.3.1 Equal Variances

We have $\sigma_1^2 = \sigma_2^2 = \sigma^2$. Define s_p, the pooled estimator of the common variance, as

$$s_p = \sqrt{\frac{(n_1 - 1)s_1^2 + (n_2 - 1)s_2^2}{n_1 + n_2 - 2}}. \tag{8.2}$$

Let $s^* = s_p\sqrt{1/n_1 + 1/n_2}$. Then it is known that (Section 7.3 of [1])

$$T = \frac{(\overline{x}_1 - \overline{x}_2) - (\mu_1 - \mu_2)}{s^*} \tag{8.3}$$

has a (Student's) t-distribution with $n_1 + n_2 - 2$ degrees of freedom. Then

$$P(-t_{\beta/2} \leq T \leq t_{\beta/2}) = 1 - \beta. \tag{8.4}$$

Solve the inequality for $\mu_1 - \mu_2$ and we find the $(1 - \beta)100\%$ confidence interval for $\mu_1 - \mu_2$

$$[\overline{x}_1 - \overline{x}_2 - t_{\beta/2}s^*, \overline{x}_1 - \overline{x}_2 + t_{\beta/2}s^*]. \tag{8.5}$$

We place these confidence intervals one on top of another, as in Chapter 3, to get our fuzzy estimator $\overline{\mu}_{12}$.

Example 8.3.1.1

Assume that the variances in the two populations are equal. Let the data be: (1) $n_1 = 15$, $\overline{x}_1 = 70.1$, $s_1^2 = 6$; and (2) $n_2 = 8$,$\overline{x}_2 = 75.3$, $s_2^2 = 4$. This data is similar to that in Example 7.1.1 in Chapter 7. We compute $s_p = 2.3094$. Then equation (8.5) becomes

$$[-5.2 - 1.0110t_{\beta/2}, -5.2 + 1.0110t_{\beta/2}]. \tag{8.6}$$

The degrees of freedom is 21. If $0.01 \leq \beta \leq 1$, then the graph of the fuzzy estimator $\overline{\mu}_{12}$, using Maple [2], is in Figure 8.1.

8.3.2 Unequal Variances

We have $\sigma_1^2 \neq \sigma_2^2$. It is known that

$$T = \frac{(\overline{x}_1 - \overline{x}_2) - (\mu_1 - \mu_1)}{s_0}, \tag{8.7}$$

is approximately (Student's) t distributed with r degrees of freedom (Section 7.3, [1]). We find the degrees of freedom (r) by rounding up to the nearest integer the following expression

$$\frac{(A + B)^2}{\frac{A^2}{n_1 - 1} + \frac{B^2}{n_2 - 1}}, \tag{8.8}$$

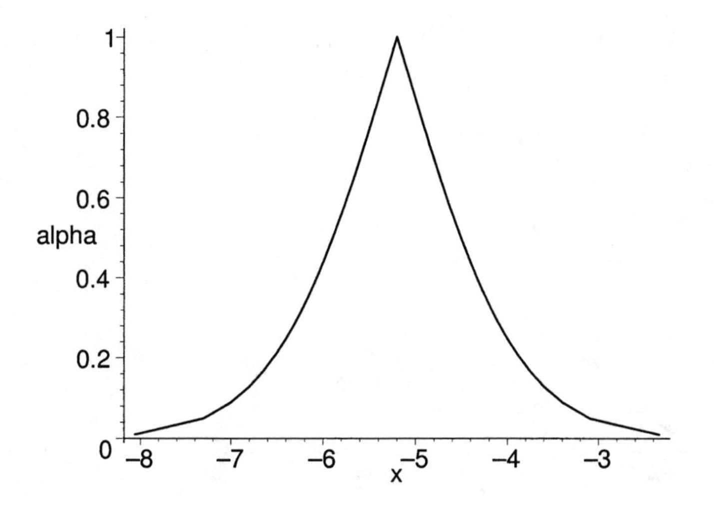

Figure 8.1: Fuzzy Estimator $\overline{\mu}_{12}$ in Example 8.3.1.1, $0.01 \leq \beta \leq 1$

where $A = s_1^2/n_1$ and $B = s_2^2/n_2$. Let

$$T = \frac{(\overline{x}_1 - \overline{x}_2) - (\mu_1 - \mu_2)}{s_0}. \tag{8.9}$$

Then

$$P(-t_{\beta/2} \leq T \leq t_{\beta/2}) \approx 1 - \beta. \tag{8.10}$$

Solve the inequality for $\mu_1 - \mu_2$ and we find an approximate $(1-\beta)100\%$ confidence interval for $\mu_1 - \mu_2$

$$[\overline{x}_1 - \overline{x}_2 - t_{\beta/2}s_0, \overline{x}_1 - \overline{x}_2 + t_{\beta/2}s_0]. \tag{8.11}$$

The t distribution has r degrees of freedom. We place these confidence intervals one on top of another, as in Chapter 3, to get our fuzzy estimator $\overline{\mu}_{12}$.

Example 8.3.2.1

Let us use the same data as in Example 8.3.1.1 except now we do not assume the variances are equal. An approximate $(1-\beta)100\%$ confidence interval for the difference of the means is

$$[-5.2 - (0.9487)t_{\beta/2}, -5.2 + (0.9487)t_{\beta/2}], \tag{8.12}$$

The term s_0 was defined in Section 8.2 and we computed it as 0.90487. Also, using equation (8.8) we determined that the degrees of freedom is $r = 18$.

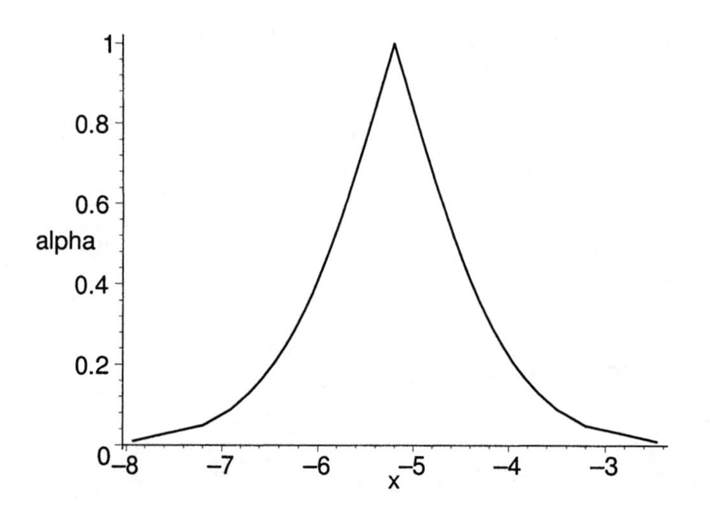

Figure 8.2: Fuzzy Estimator $\overline{\mu}_{12}$ in Example 8.3.2.1, $0.01 \leq \beta \leq 1$

The graph of the fuzzy estimator $\overline{\mu}_{12}$, using Maple [2], is in Figure 8.2. The graphs of $\overline{\mu}_{12}$ in Figures 8.1 and 8.2 are almost identical.

8.4 References

1. R.V.Hogg and E.A.Tanis: Probability and Statistical Inference, Sixth Edition, Prentice hall, Upper Saddle River, N.J., 2001.

2. Maple 6, Waterloo Maple Inc., Waterloo, Canada.

Chapter 9

Estimate $d = \mu_1 - \mu_2$, Matched Pairs

9.1 Fuzzy Estimator

Let $x_1, ..., x_n$ be the values of a random sample from a population Pop I. Let $object_i$, or $person_i$, belong to Pop I which produced measurement x_i in the random sample, $1 \leq i \leq n$. Then, at possibly some later time, we take a second measurement on $object_i$ ($person_i$) and get value y_i, $1 \leq i \leq n$. Then (x_1, y_1),...,(x_n, y_n) are n pairs of dependent measurements. For example, when testing the effectiveness of some treatment for high blood pressure, the x_i are the blood pressure measurements before treatment and the y_i are these measurements on the same person after treatment. The two samples are not independent so we can not use the results of Chapters 7 or 8.

Let $d_i = x_i - y_i$, $1 \leq i \leq n$. Next compute the mean $\overline{d}$ (crisp number, not fuzzy) and the variance s_d^2 of the d_i data. Assume that $n > 30$ so we may use the normal approximation; or assume that the d_i are approximately normally distributed with unknown mean μ_d and unknown variance σ_d^2. Then the statistic (Section 7.3 in [1])

$$T = \frac{\overline{d} - \mu_d}{s_d/\sqrt{n}}, \tag{9.1}$$

has a t distribution with $n - 1$ degrees of freedom. It follows that

$$P(-t_{\beta/2} \leq T \leq t_{\beta/2}) = 1 - \beta. \tag{9.2}$$

From this it immediately follows that a $(1 - \beta)100\%$ confidence interval for μ_d is

$$[\overline{d} - t_{\beta/2}\frac{s_d}{\sqrt{n}}, \overline{d} + t_{\beta/2}\frac{s_d}{\sqrt{n}}]. \tag{9.3}$$

Forecast (x)	Actual High (y)
68	72
76	74
66	62
72	76
76	75
80	78
71	75

Table 9.1: Forecast High Temperatures and Actual Values

Now place these confidence intervals given in equation (9.3) one on top of another to produce our fuzzy estimator $\overline{\mu}_d$ of μ_d.

Example 9.1.1

Consider the paired data in Table 9.1. This table contains a weeks forecast high temperatures and the actual recorded high values. We compute $d_i = x_i - y_i$, $1 \leq i \leq 7$ and then $\overline{d} = -0.4286$ and $s_d = 3.4572$.

The $(1-\beta)100\%$ confidence intervals are

$$[-0.4286 - (1.3067)t_{\beta/2}, -0.4286 + (1.3067)t_{\beta/2}]. \tag{9.4}$$

Using Maple [2] we graphed the confidence intervals for $0.01 \leq \beta \leq 1$ and the result is our fuzzy estimator $\overline{\mu}_d$ of μ_d in Figure 9.1. How accurate was the forecaster?

9.2 References

1. R.V.Hogg and E.A.Tanis: Probability and Statistical Inference, Sixth Edition, Prentice Hall, Upper Saddle River, N.J., 2001.

2. Maple 6, Waterloo Maple Inc., Waterloo, Canada.

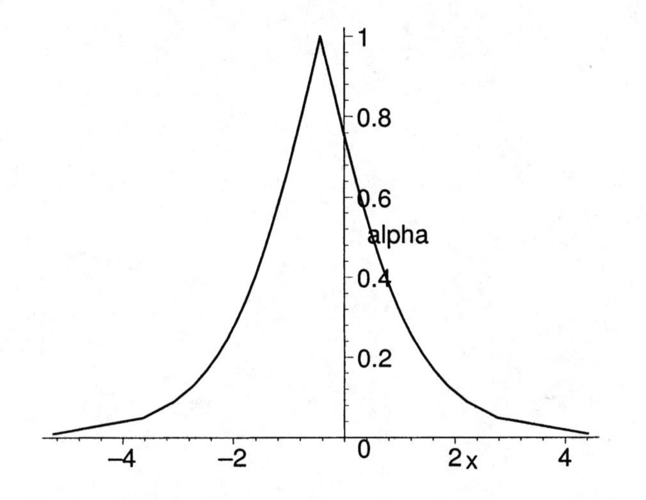

Figure 9.1: Fuzzy Estimator $\overline{\mu}_d$ in Example 9.1.1, $0.01 \leq \beta \leq 1$

Chapter 10

Estimate $p_1 - p_2$, Binomial Populations

10.1 Fuzzy Estimator

In this chapter we have two binomial population Pop I and Pop II. In Pop I (II) let p_1 (p_2) be the probability of a "success". We want a fuzzy estimator for $p_1 - p_2$.

We take a random sample of size n_1 (n_2) from Pop I (II) and observe x_1 (x_2) successes. Then our point estimator for p_1 (p_2) is $\widehat{p}_1 = x_1/n_1$ ($\widehat{p}_2 = x_2/n_2$). We assume that these two random samples are independent. Then our point estimator of $p_1 - p_2$ is $\widehat{p}_1 - \widehat{p}_2$.

Now we would like to use the normal approximation to the binomial to construct confidence intervals for $p_1 - p_2$. To do this n_1 and n_2 need to be sufficiently large. So we assume that the sample sizes are sufficiently large so that we may use the normal approximation.

Now $\widehat{p}_i$ is (approximately) normally distributed with mean p_i and variance $p_i(1-p_i)/n_i$, $i = 1, 2$. Then $\widehat{p}_1 - \widehat{p}_2$ is (approximately) normally distributed with mean $p_1 - p_2$ and variance $p_1(1-p_1)/n_1 + p_2(1-p_2)/n_2$. This would lead directly to confidence interval, but however we can not evaluate the variance expression because we do not know a value for p_1 and p_2. We solve this problem by substituting $\widehat{p}_i$ for p_i, $i = 1, 2$, in the variance equation and use $\widehat{q}_i = 1 - \widehat{p}_i$. Let

$$s_0 = \sqrt{\widehat{p}_1\widehat{q}_1/n_1 + \widehat{p}_2\widehat{q}_2/n_2}. \tag{10.1}$$

then (Section 7.5 of [1])

$$P(-z_{\beta/2} \leq \frac{(\widehat{p}_1 - \widehat{p}_2) - (p_1 - p_2)}{s_0} \leq z_{\beta/2}) \approx 1 - \beta. \tag{10.2}$$

Solve the inequality for $p_1 - p_2$ we obtain an approximate $(1-\beta)100\%$ con-

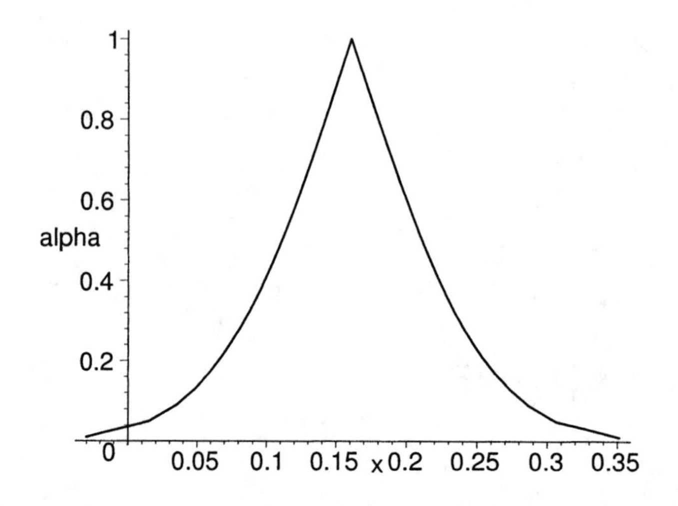

Figure 10.1: Fuzzy Estimator $\overline{p}_{12}$ in Example 10.1.1, $0.01 \leq \beta \leq 1$

fidence interval for $p_1 - p_2$ as

$$[\widehat{p}_1 - \widehat{p}_2 - z_{\beta/2}s_0, \widehat{p}_1 - \widehat{p}_2 + z_{\beta/2}s_0]. \tag{10.3}$$

Put these confidence intervals together to produce our fuzzy estimator $\overline{p}_{12}$ for $p_1 - p_2$.

Example 10.1.1

Let the data be: (1) $x_1 = 63$, $n_1 = 91$; and (2) $x_2 = 42$, $n_2 = 79$. Then equation (10.3) becomes

$$[0.1607 - 0.0741z_{\beta/2}, 0.1607 + 0.0741z_{\beta/2}]. \tag{10.4}$$

To obtain a graph of $\overline{p}_{12}$ assume that $0.01 \leq \beta \leq 1$. We evaluated equation (10.3) using Maple [2] and then the graph of $\overline{p}_{12}$ is shown in Figure 10.1, without dropping the graph straight down to the x-axis at the end points.

10.2 References

1. R.V.Hogg and E.A.Tanis: Probability and Statistical Inference, Sixth Edition, Prentice Hall, Upper Saddle River, N.J., 2001.

2. Maple 6, Waterloo Maple Inc., Waterloo, Canada.

Chapter 11

Estimate σ_1^2/σ_2^2, Normal Populations

11.1 Introduction

We first discuss the non-fuzzy confidence intervals for the ratio of the variances. We then construct a fuzzy estimator for σ_1^2/σ_2^2 from these crisp confidence intervals and show it is biased (see Chapter 6). Then we find an unbiased fuzzy estimator.

11.2 Crisp Estimator

There are two normal populations Pop I and Pop II where: (1) Pop I is $N(\mu_1, \sigma_1^2)$; and (2) Pop II is $N(\mu_1, \sigma_2^2)$. We want to get confidence intervals for σ_1^2/σ_2^2. To estimate σ_1^2 (σ_2^2) we obtain a random sample of size n_1 (n_2) from Pop I (Pop II) and compute s_1^2 (s_2^2) the sample variance (equation (6.1) in Chapter 6). Assume the two random samples were independent. Then we know (Section 7.4 in [1]) that

$$f_0 = (s_2^2/\sigma_2^2)/(s_1^2/\sigma_1^2), \tag{11.1}$$

has a F distribution with $n_2 - 1$ degrees of freedom (numerator) and $n_1 - 1$ degrees of freedom (denominator). From the F distribution we find constants a and b so that

$$P(a \leq f_0 \leq b) = 1 - \beta. \tag{11.2}$$

Then

$$P(a\frac{s_1^2}{s_2^2} \leq \frac{\sigma_1^2}{\sigma_2^2} \leq b\frac{s_1^2}{s_2^2}) = 1 - \beta. \tag{11.3}$$

It immediately follows that a $(1-\beta)100\%$ confidence interval for σ_1^2/σ_2^2 is

$$[a(s_1^2/s_2^2), b(s_1^2/s_2^2)]. \tag{11.4}$$

Now to determine the a and b.

Assume that X is a random variable from a F distribution with degrees of freedom u (numerator) and v (denominator). Let $F_{L,\beta/2}(u,v)$ be a constant so that $P(X \leq F_{L,\beta/2}(u,v)) = \beta/2$. Also let $F_{R,\beta/2}(u,v)$ be another constant so that $P(X \geq F_{R,\beta/2}(u,v)) = \beta/2$. Then the usual confidence interval has $a = F_{L,\beta/2}(u,v)$ and $b = F_{R,\beta/2}(u,v)$ which gives

$$[F_{L,\beta/2}(n_2-1,n_1-1)\frac{s_1^2}{s_2^2}, F_{R,\beta/2}(n_2-1,n_1-1)\frac{s_1^2}{s_2^2}], \tag{11.5}$$

as the $(1-\beta)100\%$ confidence interval for σ_1^2/σ_2^2. A $(1-\beta)100\%$ confidence interval for σ_1/σ_2 would be

$$[\sqrt{F_{L,\beta/2}(n_2-1,n_1-1)}(s_1/s_2), \sqrt{F_{R,\beta/2}(n_2-1,n_1-1)}(s_1/s_2)]. \tag{11.6}$$

11.3 Fuzzy Estimator

Our fuzzy estimator of σ_1^2/σ_2^2 would be constructed by placing the confidence intervals in equation (11.5) one on top of another. However, this fuzzy estimator is biased. It is biased because the vertex (membership value one) is not at the point estimator s_1^2/s_2^2. To obtain the value at the vertex we substitute one for β and get the 0% confidence interval $[c(s_1^2/s_2^2), c(s_1^2/s_2^2)] = c(s_1^2/s_2^2)$ where $c = F_{L,0.5}(n_2-1,n_1-1) = F_{R,0.5}(n_2-1,n_1-1)$. Usually the constant $c \neq 1$. We will have $c = 1$ if $n_1 = n_2$. Since c is usually not one the 0% confidence interval will not always be the point estimator. Let us now build an unbiased fuzzy estimator for σ_1^2/σ_2^2.

Our method of making an unbiased fuzzy estimator is similar to what we did in Chapter 6. Assume that $0.01 \leq \beta \leq 1$. Now this interval for β is fixed and also n_1, n_2, s_1^2 and s_2^2 are fixed. Define

$$L(\lambda) = [1-\lambda]F_{L,0.005}(n_2-1,n_1-1) + \lambda, \tag{11.7}$$

and

$$R(\lambda) = [1-\lambda]F_{R,0.005}(n_2-1,n_1-1) + \lambda. \tag{11.8}$$

The $L(\lambda)$ $(R(\lambda))$ defined above are different from the $L(\lambda)$ $(R(\lambda))$ defined in equation (6.8) ((6.9)) in Chapter 6. Then a confidence interval for the ratio of the variances is

$$[L(\lambda)\frac{s_1^2}{s_2^2}, R(\lambda)\frac{s_1^2}{s_2^2}], \tag{11.9}$$

for $0 \leq \lambda \leq 1$. We start with a 99% confidence interval when $\lambda = 0$ and end up with a 0% confidence interval for $\lambda = 1$. $L(\lambda)$ $(R(\lambda))$ continuously increases (decreases) to one as λ goes from zero to one. Notice that now the 0% confidence interval is $[s_1^2/s_2^2, s_1^2/s_2^2] = s_1^2/s_2^2$ and it is unbiased. As usual, we place these confidence intervals one on top of another to obtain our (unbiased) fuzzy estimator $\overline{\sigma}_{12}^2$ for the ratio of the variances. Our confidence interval for σ_1/σ_2, the ratio of the population standard deviations, is

$$[\sqrt{L(\lambda)}(s_1/s_2), \sqrt{R(\lambda)}(s_1/s_2)]. \tag{11.10}$$

These confidence intervals will make up our fuzzy estimator $\overline{\sigma}_{12}$ for σ_1/σ_2. We may find the relationship between λ and β because β is a function of λ given by

$$\beta = \int_0^{L(\lambda)} F dx + \int_{R(\lambda)}^{\infty} F dx, \tag{11.11}$$

where "F" denotes the F distribution with $n_2 - 1$ and $n_1 - 1$ degrees of freedom.

Example 11.3.1

From Pop I we have a random sample of size $n_1 = 8$ and we compute $s_1^2 = 14.3$. From Pop II the data was $n_2 = 12$ and $s_2^2 = 9.8$. Then

$$L(\lambda) = (1 - \lambda)(0.1705) + \lambda, \tag{11.12}$$

$$R(\lambda) = (1 - \lambda)(8.2697) + \lambda. \tag{11.13}$$

The confidence intervals become

$$[(0.1705 + 0.8295\lambda)(1.459), (8.2697 - 7.2697\lambda)(1.459)], \tag{11.14}$$

for $0 \leq \lambda \leq 1$. Maple [2] produced the graph of $\overline{\sigma}_{12}^2$ in Figure 11.1 from equation (11.14). Notice that in this case we obtain a triangular fuzzy number (sides are straight line segments). Figure 11.2 has the graph of $\overline{\sigma}_{12}$. The Maple commands for Figure 11.1 are in Chapter 29.

11.4 References

1. R.V.Hogg and E.A.Tanis: Probability and Statistical Inference, Sixth Edition, Prentice hall, Upper Saddle River, N.J., 2001.

2. Maple 6, Waterloo Maple Inc., Waterloo, Canada.

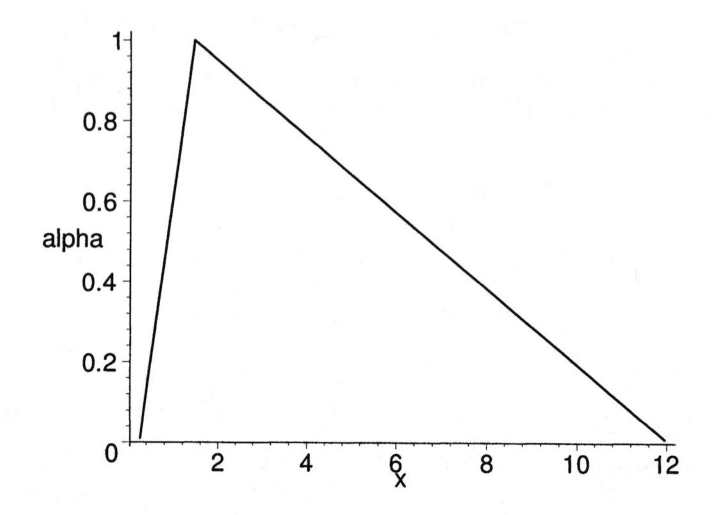

Figure 11.1: Fuzzy Estimator $\overline{\sigma}_{12}^2$ of σ_1^2/σ_2^2 in Example 11.3.1, $0.01 \leq \beta \leq 1$

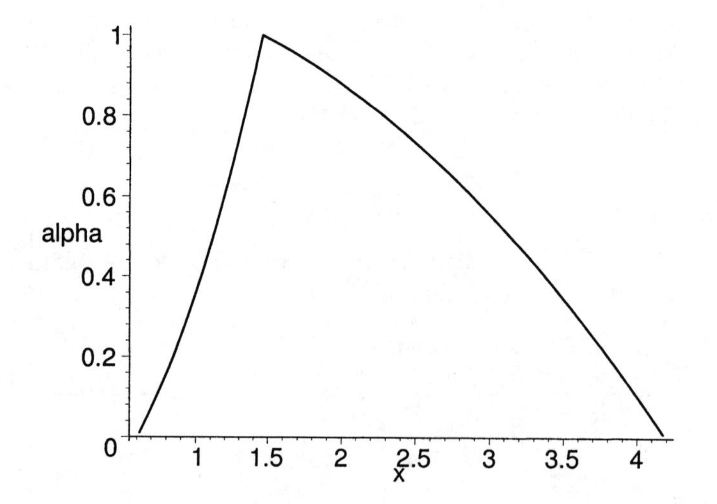

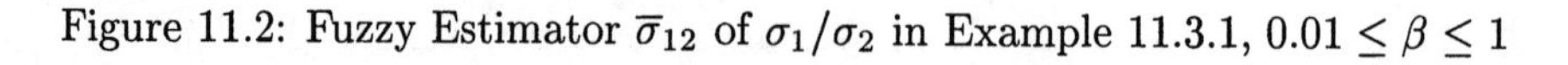

Figure 11.2: Fuzzy Estimator $\overline{\sigma}_{12}$ of σ_1/σ_2 in Example 11.3.1, $0.01 \leq \beta \leq 1$

Chapter 12

Tests on μ, Variance Known

12.1 Introduction

This chapter starts a series of chapters (Chapters 12-20) on fuzzy hypothesis testing. In each chapter we first review the crisp case first before proceeding to the fuzzy situation. We give more details on fuzzy hypothesis testing in this chapter.

In the previous chapters on estimation , Chapters 3-11, we sometimes gave multiple graphs of the fuzzy estimators, like for $0.10 \leq \beta \leq 1$ and $0.01 \leq \beta \leq 1$ and $0.001 \leq \beta \leq 1$. We shall not do this again in this book. Unless specified differently, we will always use $0.01 \leq \beta \leq 1$ in the rest of this book. Also, in these previous chapters on estimation we kept reminding the reader about those very short vertical line segments at the two ends of any fuzzy estimator. We will mention this fact only a couple of more times in the rest of this book.

12.2 Non-Fuzzy Case

We obtain a random sample of size n from a $N(\mu, \sigma^2)$, variance σ^2 known, in order to do the following hypothesis test

$$H_0 : \mu = \mu_0, \tag{12.1}$$

verses

$$H_1 : \mu \neq \mu_0. \tag{12.2}$$

In this book we will usually start with the alternate hypothesis H_1 two-sided ($\mu \neq \mu_0$) instead of one-sided ($\mu > \mu_0$ or $\mu < \mu_0$). Using a two-sided alternate hypothesis makes the discussion a little more general, and at the end of this chapter we will show what changes need to be made for the one-sided tests. From the random sample we compute its mean $\overline{x}$ (a real number), and then determine the statistic

$$z_0 = \frac{\overline{x} - \mu_0}{\sigma/\sqrt{n}}. \tag{12.3}$$

Let γ, $0 < \gamma < 1$, be the significance level of the test. Usual values for γ are $0.10, 0.05, 0.01$. Now under the null hypothesis H_0 z_0 is $N(0,1)$ (Section 8.2 in [1]) and our decision rule is: (1) reject H_0 if $z_0 \geq z_{\gamma/2}$ or $z_0 \leq -z_{\gamma/2}$; and (2) do not reject H_0 when $-z_{\gamma/2} < z_0 < z_{\gamma/2}$. The numbers $\pm z_{\gamma/2}$ are called the critical values (cv's) for the test. In the above decision rule $z_{\gamma/2}$ is the z-value so that the probability of a random variable, having the $N(0,1)$ probability density, exceeding z is $\gamma/2$. Usually authors use α for the significance level of a test but in this book we will use α for α-cuts of fuzzy numbers.

12.3 Fuzzy Case

Now proceed to the fuzzy situation where our estimate of μ, as explained in Chapter 3, is the triangular shaped fuzzy number $\overline{\mu}$ where its α-cuts are

$$\overline{\mu}[\alpha] = [\overline{x} - z_{\alpha/2}\sigma/\sqrt{n}, \overline{x} + z_{\alpha/2}\sigma/\sqrt{n}], \tag{12.4}$$

for $0.01 \leq \alpha \leq 1$. In the rest of the book we will always have the base of the fuzzy estimator a 99% confidence interval. Recall that the alpha-cuts of $\overline{\mu}$, for $0 \leq \alpha \leq 0.01$, all equal $\overline{\mu}[0.01]$.

Calculations will be performed by alpha-cuts and interval arithmetic (Sections 2.3.2 and 2.3.3). Our fuzzy statistic becomes

$$\overline{Z} = \frac{\overline{\mu} - \mu_0}{\sigma/\sqrt{n}}. \tag{12.5}$$

Now substitute alpha-cuts of $\overline{\mu}$, equation (12.4), into equation (12.5) and simplify using interval arithmetic producing alpha-cuts of $\overline{Z}$

$$\overline{Z}[\alpha] = [z_0 - z_{\alpha/2}, z_0 + z_{\alpha/2}]. \tag{12.6}$$

We put these α-cuts together to get a fuzzy triangular shaped fuzzy number $\overline{Z}$.

Since our test statistic is fuzzy the critical values will also be fuzzy. There will be two fuzzy critical value sets: (1) let $\overline{CV}_1$ correspond to $-z_{\gamma/2}$; and (2) let $\overline{CV}_2$ go with $z_{\gamma/2}$. Set $\overline{CV}_i[\alpha] = [cv_{i1}(\alpha), cv_{i2}(\alpha)]$, $i = 1, 2$. We show how to get $cv_{21}(\alpha)$ and $cv_{22}(\alpha)$. The end points of an alpha-cut of $\overline{CV}_2$ are computed from the end points of the corresponding alpha-cut of $\overline{Z}$. We see that to find $cv_{22}(\alpha)$ we solve

$$P(z_0 + z_{\alpha/2} \geq cv_{22}(\alpha)) = \gamma/2, \tag{12.7}$$

for $cv_{22}(\alpha)$. The above equation is the same as

$$P(z_0 \geq cv_{22}(\alpha) - z_{\alpha/2}) = \gamma/2. \tag{12.8}$$

But z_0 is $N(0,1)$ so

$$cv_{22}(\alpha) - z_{\alpha/2} = z_{\gamma/2}, \tag{12.9}$$

or

$$c_{22}(\alpha) = z_{\gamma/2} + z_{\alpha/2}. \tag{12.10}$$

By using the left end point of $\overline{Z}[\alpha]$ in equation (12.7) we have

$$cv_{21}(\alpha) = z_{\gamma/2} - z_{\alpha/2}. \tag{12.11}$$

Hence an alpha-cut of $\overline{CV}_2$ is

$$[z_{\gamma/2} - z_{\alpha/2}, z_{\gamma/2} + z_{\alpha/2}]. \tag{12.12}$$

In the above equation for $\overline{CV}_2[\alpha]$, γ is fixed, and α ranges in the interval $[0.01, 1]$. Now $\overline{CV}_1 = -\overline{CV}_2$ so

$$\overline{CV}_1[\alpha] = [-z_{\gamma/2} - z_{\alpha/2}, -z_{\gamma/2} + z_{\alpha/2}]. \tag{12.13}$$

Both $\overline{CV}_1$ and $\overline{CV}_2$ will be triangular shaped fuzzy numbers. When the crisp test statistic has a normal, or t, distribution we will have $\overline{CV}_1 = -\overline{CV}_2$ because these densities are symmetric with respect to zero. However, if the crisp test statistic has the χ^2, or the F, distribution we will have $\overline{CV}_1 \neq -\overline{CV}_2$ because these densities are not symmetric with respect to zero.

Let us present another derivation of $\overline{CV}_2$. Let $0.01 \leq \alpha < 1$ and choose $z \in \overline{Z}[\alpha]$. This value of z is a possible value of the crisp test statistic corresponding to the $(1-\alpha)100\%$ confidence interval for μ. Then the critical value cv_2 corresponding to z belongs to $\overline{CV}_2[\alpha]$. In fact, as z ranges throughout the interval $\overline{Z}[\alpha]$ its corresponding cv_2 will range throughout the interval $\overline{CV}_2[\alpha]$. Let $z = \tau(z_0 - z_{\alpha/2}) + (1-\tau)(z_0 + z_{\alpha/2})$ for some τ in $[0,1]$. Then

$$P(z \geq cv_2) = \gamma/2. \tag{12.14}$$

It follows that

$$P(z_0 \geq cv_2 + (2\tau - 1)z_{\alpha/2}) = \gamma/2. \tag{12.15}$$

Since z_0 is $N(0,1)$ we obtain

$$cv_2 + (2\tau - 1)z_{\alpha/2} = z_{\gamma/2}, \tag{12.16}$$

or

$$cv_2 = \tau(z_{\gamma/2} - z_{\alpha/2}) + (1-\tau)(z_{\gamma/2} + z_{\alpha/2}), \tag{12.17}$$

which implies that $cv_2 \in \overline{CV}_2[\alpha]$. This same argument can be given in the rest of the book whenever we derive the $\overline{CV}_i$, $i = 1, 2$, however, we shall not go through these details again.

Our final decision (reject, do not reject) will depend on the relationship between $\overline{Z}$ and the $\overline{CV}_i$. Now it is time to review Section 2.5 and Figure 2.4 of Chapter 2. In comparing $\overline{Z}$ and $\overline{CV}_1$ we will obtain $\overline{Z} < \overline{CV}_1$ (reject H_0), or $\overline{Z} \approx \overline{CV}_1$(no decision), or $\overline{Z} > \overline{CV}_1$ (do not reject). Similar results when comparing $\overline{Z}$ and $\overline{CV}_2$. Let R stand for "reject" H_0, let DNR be "do not reject" H_0 and set ND to be "no decision". After comparing $\overline{Z}$ and the $\overline{CV}_i$ we get (A, B) for $A, B \in \{R, DNR, ND\}$ where A (B) is the result of $\overline{Z}$ verses $\overline{CV}_1$ $(\overline{CV}_2)$. We suggest the final decision to be: (1) if A or B is R, then "reject" H_0; (2) if A and B are both DNR, then "do not reject" H_0, (3) if both A and B are ND, then we have "no decision"; and (4) if $(A, B) = (ND, DNR)$ or (DNR, ND), then "no decision". The only part of the above decision rule that might be debatable is the fourth one. Users may wish to change the fourth one to "do not reject". However, the author prefers "no decision".

It is interesting that in the fuzzy case we can end up in the "no decision" case. This is because of the fuzzy numbers, which incorporate all the uncertainty in the confidence intervals, that we can get $\overline{M} \approx \overline{N}$ for two fuzzy numbers $\overline{M}$ and $\overline{N}$.

Let us go through some more details on deciding $\overline{Z} <, \approx, > \overline{CV}_i$ using Section 2.5 and Figure 2.4 before going on to examples. First consider $\overline{Z}$ verses $\overline{CV}_2$. We may have $z_0 > z_{\gamma/2}$ or $z_0 = z_{\gamma/2}$ or $z_0 < z_{\gamma/2}$. First consider $z_0 > z_{\gamma/2}$. So draw $\overline{Z}$ to the right of $\overline{CV}_2$ and find the height of the intersection (the left side of $\overline{Z}$ with the right side of $\overline{CV}_2$). Let the height of the intersection be y_0 as in Section 2.5. If there is no intersection then set $y_0 = 0$. Recall that we are using the test number $\eta = 0.8$ from Section 2.5. The results are: (1) if $y_0 < 0.8$, then $\overline{Z} > \overline{CV}_2$; and (2) if $y_0 \geq 0.8$, then $\overline{Z} \approx \overline{CV}_2$. If $z_0 = z_{\gamma/2}$ then $\overline{Z} \approx \overline{CV}_2$. So now assume that $z_0 < z_{\gamma/2}$. The height of the intersection y_0 will be as shown in Figure 2.4. The decision is: (1) if $y_0 < 0.8$, then $\overline{Z} < \overline{CV}_2$; and (2) if $y_0 \geq 0.8$, then $\overline{Z} \approx \overline{CV}_2$. Similar results hold for $\overline{Z}$ verses $\overline{CV}_1$.

A summary of the cases we expect to happen are: (1) $\overline{CV}_2 < \overline{Z}$ reject; (1) $\overline{CV}_1 < \overline{Z} \approx \overline{CV}_2$ no decision; (3) $\overline{CV}_1 < \overline{Z} < \overline{CV}_2$ do not reject; (4) $\overline{CV}_1 \approx \overline{Z} < \overline{CV}_2$ no decision; and (5) $\overline{Z} < \overline{CV}_1$ reject.

Example 12.3.1

Assume that $n = 100$, $\mu_0 = 1$, $\sigma = 2$ and the significance level of the test is $\gamma = 0.05$ so $z_{\gamma/2} = 1.96$. From the random sample let $\overline{x} = 1.32$ and we then compute $z_0 = 1.60$. Recall that we will be using $0.01 \leq \alpha \leq 1$ in these chapters on fuzzy hypothesis testing.

Since $z_0 < z_{\gamma/2}$ we need to compare the right side of $\overline{Z}$ to the left side of $\overline{CV}_2$. These are shown in Figure 12.1. The Maple [2] commands for Figure 12.1 are in Chapter 29. The right side of $\overline{Z}$ decreases form $(1.60, 1)$ towards zero and the left side of $\overline{CV}_2$ increases from just above zero to $(1.96, 1)$. Draw

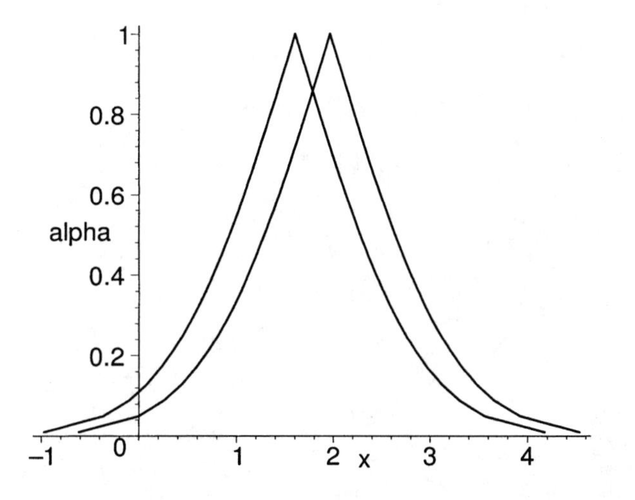

Figure 12.1: Fuzzy Test $\overline{Z}$ verses $\overline{CV}_2$ in Example 12.3.1($\overline{Z}$ left, $\overline{CV}_2$ right)

a horizontal line through 0.8 on the vertical axis and we observe that the height of intersection is greater than 0.8. From this comparison we conclude $\overline{Z} \approx \overline{CV}_2$.

Next we evaluate $\overline{Z}$ verses $\overline{CV}_1$. The results are in Figure 12.2. Since the height of the intersection is less than 0.8 we conclude $\overline{CV}_1 < \overline{Z}$. Our final conclusion is no decision on H_0. In the crisp case since $-z_{\gamma/2} < z_0 < z_{\gamma/2}$ we would decide: do not reject H_0.

Example 12.3.2

All the data is the same as in Example 12.3.1 except now assume that $\overline{x} = 0.40$. We compute $z_0 = -3.0$. Since $z_0 < -z_{\gamma/2}$ we first compare $\overline{Z}$ and $\overline{CV}_1$. This is shown in Figure 12.3. The right side of $\overline{Z}$ decreases from $(-3.0, 1)$ towards zero and the left side of $\overline{CV}_1$ increases from near zero to $(-1.96, 1)$. The height of the intersection is clearly less than 0.8 so $\overline{Z} < \overline{CV}_1$. It is obvious that we also have $\overline{Z} < \overline{CV}_2$. Hence we reject H_0. In the crisp case we would also reject H_0.

12.4 One-Sided Tests

First consider

$$H_0 : \mu = \mu_0, \tag{12.18}$$

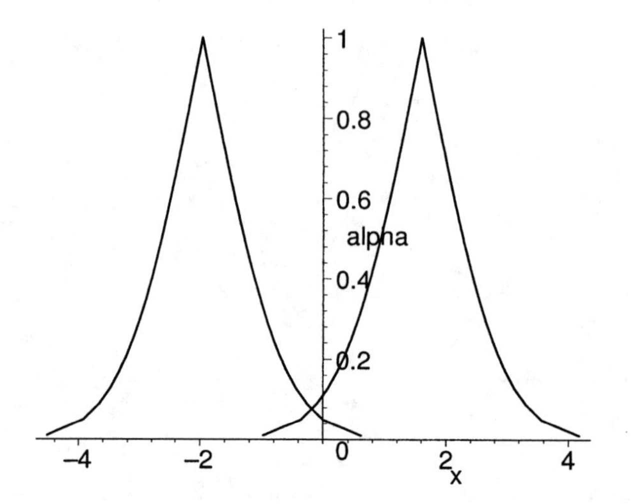

Figure 12.2: Fuzzy Test $\overline{Z}$ verses $\overline{CV}_1$ in Example 12.3.1($\overline{CV}_1$ left, $\overline{Z}$ right)

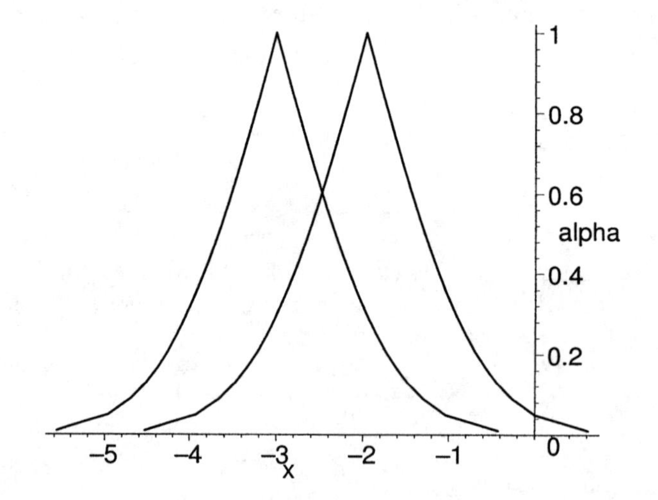

Figure 12.3: Fuzzy Test $\overline{Z}$ verses $\overline{CV}_1$ in Example 12.3.2($\overline{Z}$ left, $\overline{CV}_1$ right)

verses

$$H_1 : \mu > \mu_0. \tag{12.19}$$

Then we would reject H_0 when $z_0 \geq z_\gamma$, here we use γ and not $\gamma/2$, and do not reject if $z_0 < z_\gamma$.

In the fuzzy case we still have the same fuzzy statistic $\overline{Z}$, no $\overline{CV}_1$, and now $\overline{CV}_2$ would be centered (membership value one) at z_γ. The final decision would be made only through the comparison of $\overline{Z}$ and $\overline{CV}_2$. The three case are: (1) if $\overline{Z} < \overline{CV}_2$, then do not reject, (2) if $\overline{Z} \approx \overline{CV}_2$, then no decision on H_0; and (3) if $\overline{Z} > \overline{CV}_2$, then reject H_0.

Obvious changes are to be made if we have the other one-sided test

$$H_0 : \mu = \mu_0, \tag{12.20}$$

verses

$$H_1 : \mu < \mu_0. \tag{12.21}$$

We shall use these one-sided fuzzy hypothesis tests in Chapters 19, 24 and 27.

12.5 References

1. R.V.Hogg and E.A.Tanis: Probability and Statistical Inference, Sixth Edition, Prentice Hall, Upper Saddle River, N.J., 2001.

2. Maple 6, Waterloo Maple Inc., Waterloo, Canada.

Chapter 13

Tests on μ, Variance Unknown

13.1 Introduction

We start with a review of the crisp case and then present the fuzzy model.

13.2 Crisp Case

We obtain a random sample of size n from a $N(\mu, \sigma^2)$, mean and variance unknown, in order to do the following hypothesis test

$$H_0 : \mu = \mu_0, \tag{13.1}$$

verses

$$H_1 : \mu \neq \mu_0. \tag{13.2}$$

From the random sample we compute its mean $\overline{x}$, this will be a real number and not a fuzzy set, and the sample variance s^2 and then determine the statistic

$$t_0 = \frac{\overline{x} - \mu_0}{s/\sqrt{n}}. \tag{13.3}$$

Let γ, $0 < \gamma < 1$, be the significance level of the test. Under the null hypothesis H_0 t_0 has a t-distribution with $n-1$ degrees of freedom (Section 8.2 of [1]) and our decision rule is: (1) reject H_0 if $t_0 \geq t_{\gamma/2}$ or $t_0 \leq -t_{\gamma/2}$; and (2) do not reject H_0 when $-t_{\gamma/2} < t_0 < t_{\gamma/2}$. The numbers $\pm t_{\gamma/2}$ are called the critical values (cv's) for the test. In the above decision rule $t_{\gamma/2}$ is the t-value so that the probability of a random variable, having a t-distribution with $n-1$ degrees of freedom, exceeding t is $\gamma/2$.

13.3 Fuzzy Model

Now proceed to the fuzzy situation where our statistic for t_0 will become fuzzy $\overline{T}$. In equation (13.3) for t_0 substitute an alpha-cut of our fuzzy estimator $\overline{\mu}$ for $\overline{x}$, equation (4.4) of Chapter 4; and substitute an alpha-cut of our fuzzy estimator $\overline{\sigma}$ for s, the square root of equation (6.10) of Chapter 6. Use interval arithmetic to simplify and we obtain

$$\overline{T}[\alpha] = [\Pi_1(t_0 - t_{\alpha/2}), \Pi_2(t_0 + t_{\alpha/2})], \tag{13.4}$$

where

$$\Pi_1 = \sqrt{\frac{R(\lambda)}{n-1}}, \tag{13.5}$$

and

$$\Pi_2 = \sqrt{\frac{L(\lambda)}{n-1}}, \tag{13.6}$$

for $L(\lambda)$ $(R(\lambda))$ defined in equation (6.8) ((6.9)) of Chapter 6. In the interval arithmetic employed above, we assumed that all intervals were positive ($[a, b] > 0$ if $a > 0$). It may happen that for certain values of alpha: (1) the interval in the numerator the left end point is negative, the other end point is positive, but the other interval in the denominator is positive; or (2) the interval in the numerator the right end point of the interval is negative, and so the other end point is also negative, and the other interval in the denominator is positive. These special cases will be discussed at the end of this section. For now we assume that all intervals are positive.

Since our test statistic is fuzzy the critical values will also be fuzzy. There will be two fuzzy critical value sets: (1) let $\overline{CV}_1$ correspond to $-t_{\gamma/2}$; and (2) let $\overline{CV}_2$ go with $t_{\gamma/2}$. Set $\overline{CV}_i[\alpha] = [cv_{i1}(\alpha), cv_{i2}(\alpha)]$, $i = 1, 2$. We show how to get $cv_{21}(\alpha)$ and $cv_{22}(\alpha)$. The end points of an alpha-cut of $\overline{CV}_2$ are computed from the end points of the corresponding alpha-cut of $\overline{T}$. We see that to find $cv_{22}(\alpha)$ we solve

$$P(\Pi_2(t_0 + t_{\alpha/2}) \geq cv_{22}(\alpha)) = \gamma/2, \tag{13.7}$$

for $cv_{22}(\alpha)$. The above equation is the same as

$$P(t_0 \geq (cv_{22}(\alpha)/\Pi_2) - t_{\alpha/2}) = \gamma/2. \tag{13.8}$$

But t_0 has a t distribution so

$$(cv_{22}(\alpha)/\Pi_2) - t_{\alpha/2} = t_{\gamma/2}, \tag{13.9}$$

or

$$c_{22}(\alpha) = \Pi_2(t_{\gamma/2} + t_{\alpha/2}). \tag{13.10}$$

By using the left end point of $\overline{T}[\alpha]$ we have

$$cv_{21}(\alpha) = \Pi_1(t_{\gamma/2} - t_{\alpha/2}). \tag{13.11}$$

Hence an alpha-cut of $\overline{CV}_2$ is

$$[\Pi_1(t_{\gamma/2} - t_{\alpha/2}), \Pi_2(t_{\gamma/2} + t_{\alpha/2})]. \tag{13.12}$$

In the above equation for $\overline{CV}_2[\alpha]$, γ is fixed, and α ranges in the interval $[0.01, 1]$. Now $\overline{CV}_1 = -\overline{CV}_2$ so

$$\overline{CV}_1[\alpha] = [\Pi_2(-t_{\gamma/2} - t_{\alpha/2}), \Pi_1(-t_{\gamma/2} + t_{\alpha/2})]. \tag{13.13}$$

Both $\overline{CV}_1$ and $\overline{CV}_2$ will be triangular shaped fuzzy numbers.

The details of comparing $\overline{T}$ to $\overline{CV}_1$ and $\overline{CV}_2$, and our method of coming to a final decision (reject, no decision, or do not reject), is all in Chapter 12. Before we can go on to work two examples we need to solve a major problem with these fuzzy numbers $\overline{T}$ and the $\overline{CV}_i$: their α-cuts depend on two variables λ and α. $R(\lambda)$ and $L(\lambda)$ are obviously functions of λ and $t_{\alpha/2}$ is a function of α. But, as pointed out in Chapter 6 $\alpha = f(\lambda)$, or α is a function of λ. This comes from equations (6.12) and (6.13) of Chapter 6

$$\alpha = f(\lambda) = \int_0^{R(\lambda)} \chi^2 dx + \int_{L(\lambda)}^{\infty} \chi^2 dx, \tag{13.14}$$

as λ goes from zero to one. The chi-square density in the integrals has $n-1$ degrees of freedom. When $\lambda = 0$ then $\alpha = 0.01$ and if $\lambda = 1$ so does $\alpha = 1$. Notice that $\overline{T}[1] = [t_0, t_0] = t_0$, $\overline{CV}_1[1] = [-t_{\gamma/2}, -t_{\gamma/2}] = -t_{\gamma/2}$ and $\overline{CV}_2[1] = [t_{\gamma/2}, t_{\gamma/2}] = t_{\gamma/2}$.

To generate the triangular shaped fuzzy number $\overline{T}$ from its α-cuts in equation (13.4) we increase λ from zero to one (this determines the Π_i), compute α from λ in equation (13.14), which gives $t_{\alpha/2}$, and we have the alpha-cuts since t_0 is a constant. Put these alpha-cuts together to have $\overline{T}$. Similar remarks for the other two triangular shaped fuzzy numbers $\overline{CV}_i$, $i = 1, 2$.

13.3.1 $\overline{T}[\alpha]$ for Non-Positive Intervals

In finding equation (13.4) we divided two intervals

$$\overline{T}[\alpha] = \frac{[a, b]}{[c, d]}, \tag{13.15}$$

where

$$a = \overline{x} - t_{\alpha/2} s/\sqrt{n} - \mu_0, \tag{13.16}$$

and

$$b = \overline{x} + t_{\alpha/2} s/\sqrt{n} - \mu_0, \tag{13.17}$$

and

$$c = \sqrt{\frac{n-1}{L(\lambda)}}(s/\sqrt{n}), \tag{13.18}$$

and

$$d = \sqrt{\frac{n-1}{R(\lambda)}}(s/\sqrt{n}). \tag{13.19}$$

Now c and d are always positive but a or b could be negative for certain values of alpha.

First assume that $t_0 > 0$ so that $b > 0$ also. Also assume that there is a value of alpha, say α^* in $(0, 1)$, so that $a < 0$ for $0.01 \leq \alpha < \alpha^*$ but $a > 0$ for $\alpha^* < \alpha \leq 1$. Then we compute, using interval arithmetic,

$$\overline{T}[\alpha] = [a, b][1/d, 1/c] = [a/c, b/c], \tag{13.20}$$

for $0.01 \leq \alpha < \alpha^*$ when $a < 0$ and

$$\overline{T}[\alpha] = [a, b][1/d, 1/c] = [a/d, b/c], \tag{13.21}$$

when $\alpha^* < \alpha \leq 1$ for $a > 0$. This case will be in the following Example 13.3.1. The $a > 0$ was what we were using above for equation (13.4). We saw in Chapter 12 that the alpha-cuts of the fuzzy statistic determines the alpha-cuts of the fuzzy critical values. In this case $\overline{T}[\alpha]$ will determine $\overline{CV}_2[\alpha]$. So if we change how we compute $\overline{T}[\alpha]$ when $a < 0$, then we need to use this to find the new α-cuts of $\overline{CV}_2$ and then $\overline{CV}_1 = -\overline{CV}_2$.

Now if $t_0 < 0$ then $a < 0$ but b may be positive for some alpha and negative for other α. So assume that $b < 0$ for $0 < \alpha^* < \alpha \leq 1$ and $b > 0$ otherwise. then

$$\overline{T}[\alpha] = [a, b][1/d, 1/c] = [a/c, b/d], \tag{13.22}$$

when $b < 0$ and

$$\overline{T}[\alpha] = [a, b][1/d, 1/c] = [a/c, b/c], \tag{13.23}$$

for $b > 0$. In Example 13.3.2 below we will be interested in the $b < 0$ case (because $t_0 < 0$). The alpha-cuts of the test statistic will determine those of the fuzzy critical values. In this case $\overline{T}[\alpha]$ determines $\overline{CV}_1[\alpha]$. When we change how we get alpha-cuts of $\overline{T}$ when $b > 0$ we use this to compute the new α-cuts of $\overline{CV}_1$ and $\overline{CV}_2 = -\overline{CV}_1$.

Assuming $a < 0$ and $b < 0$ (at least for $0 < \alpha^* < \alpha \leq 1$) since $t_0 < 0$, then

$$\overline{T}[\alpha] = [\Pi_2(t_0 - t_{\alpha/2}), \Pi_1(t_0 + t_{\alpha/2})], \tag{13.24}$$

and

$$\overline{CV}_1[\alpha] = [\Pi_2(-t_{\gamma/2} - t_{\alpha/2}), \Pi_1(-t_{\gamma/2} + t_{\alpha/2})], \tag{13.25}$$

and $\overline{CV}_2 = -\overline{CV}_1$ so that

$$\overline{CV}_2[\alpha] = [\Pi_1(t_{\gamma/2} - t_{\alpha/2}), \Pi_2(t_{\gamma/2} + t_{\alpha/2})], \tag{13.26}$$

These equations will be used in the second example below.

Example 13.3.1

Assume that the random sample size is $n = 101$ and $\mu_0 = 1$, $\gamma = 0.01$ so that $t_{\gamma/2} = 2.626$ with 100 degrees of freedom. From the data suppose we found $\overline{x} = 1.32$ and the sample variance is $s^2 = 4.04$. From this we compute $t_0 = 1.60$. In order to find the Π_i we obtain $\chi^2_{R,0.005} = 140.169$ and $\chi^2_{L,0.005} = 67.328$. Then $L(\lambda) = 140.169 - 40.169\lambda$ and $R(\lambda) = 67.328 + 32.672\lambda$ and

$$\Pi_2 = \sqrt{1.40169 - 0.40169\lambda}, \tag{13.27}$$

and

$$\Pi_1 = \sqrt{0.67329 + 0.32672\lambda}. \tag{13.28}$$

Using $\alpha = f(\lambda)$ from equation (13.14) we may now have Maple [2] do the graphs of $\overline{T}$ and the $\overline{CV}_i$. Notice that in equations (13.4),(13.12) and (13.13) for the alpha-cuts instead of $t_{\alpha/2}$ we use $t_{f(\lambda)/2}$.

In comparing $\overline{T}$ and $\overline{CV}_2$ we can see the result in Figure 13.1. The Maple commands for this figure are in Chapter 29. Since $t_0 < t_{\gamma/2}$ we only need to compare the right side of $\overline{T}$ to the left side of $\overline{CV}_2$. The height of the intersection is $y_0 > 0.8$ (the point of intersection is close to, by just greater than 0.8)and we conclude that $\overline{T} \approx \overline{CV}_2$ with no decision.

The graphs in Figure 13.1 are correct only to the right of the vertical axis. The left side of $\overline{T}$ goes negative, and we should make the adjustment described in Subsection 12.3.1 to both $\overline{T}$ and $\overline{CV}_2$, but we did not because it will not effect our conclusion.

Next we compare $\overline{T}$ to $\overline{CV}_1$. We need to take into consideration here that the left side of $\overline{T}$ will go negative (see subsection on this topic just preceding this example). Noting this fact we find that the height of the intersection (the left side of $\overline{T}$ with the right side of $\overline{CV}_1$) is less than 0.8. Hence $\overline{CV}_1 < \overline{T}$ with do not reject. It is easy to see the relationship between $\overline{T}$ and $\overline{CV}_1$ because $\overline{CV}_1 = -\overline{CV}_2$.

Our final conclusion is no decision on H_0. In the crisp case the conclusion would be do not reject since $-t_{\gamma/2} < t_0 < t_{\gamma/2}$.

Example 13.3.2

Assume the data is the same as in Example 13.3.1 except $\overline{x} = 0.74$ so that $t_0 = -1.3$. Now we use the results for $a < 0$ and $b < 0$ (for $0 < \alpha^* < \alpha \leq 1$) discussed in the Subsection 13.3.1 above. Since $-t_{\gamma/2} < t_0 < 0$ we first compare the left side of $\overline{T}$ to the right side of $\overline{CV}_1$. We see that the height of the intersection is close to, but less than, 0.8. Hence $\overline{CV}_1 < \overline{T}$ and we do not reject H_0. Again, the graph of $\overline{T}$ in Figure 13.2 is not accurate to the right of the vertical axis, which also effects the right side of $\overline{CV}_1$, but this does not change the conclusion.

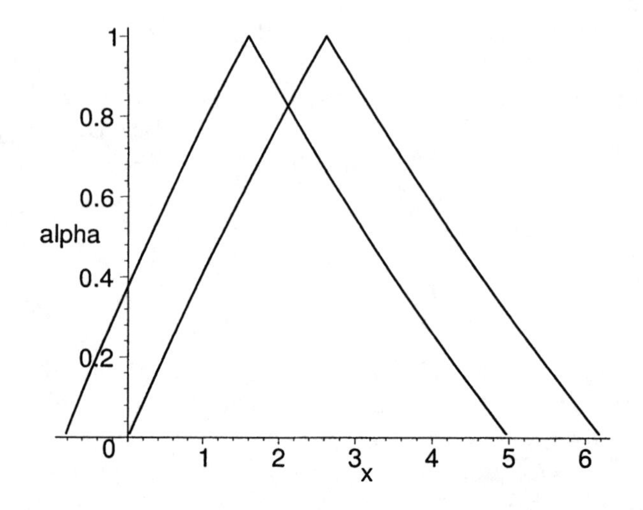

Figure 13.1: Fuzzy Test $\overline{T}$ verses $\overline{CV}_2$ in Example 13.3.1($\overline{T}$ left, $\overline{CV}_2$ right)

Next we compare the right side of $\overline{T}$, taking into account that now b will become positive, with the left side of $\overline{CV}_2$. We determine that $\overline{T} < \overline{CV}_2$ since $\overline{CV}_2 = -\overline{CV}_1$ and we do not reject H_0.

Out final conclusion is to not reject H_0. In the crisp case we also do not reject.

13.4 References

1. R.V.Hogg and E.A.Tanis: Probability and Statistical Inference, Sixth Edition, Prentice Hall, Upper Saddle River, N.J., 2001.

2. Maple 6, Waterloo Maple Inc., Waterloo, Canada.

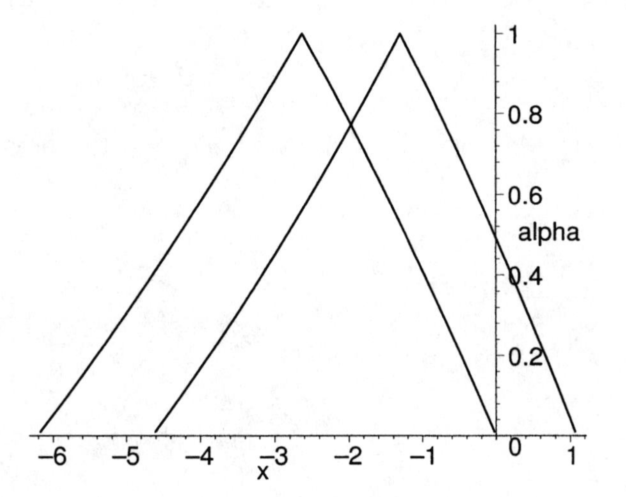

Figure 13.2: Fuzzy Test $\overline{T}$ verses $\overline{CV}_1$ in Example 13.3.2($\overline{T}$ right, $\overline{CV}_1$ left)

Chapter 14

Tests on p for a Binomial Population

14.1 Introduction

We begin with the non-fuzzy test and then proceed to the fuzzy test.

14.2 Non-Fuzzy Test

We obtain a random sample of size n from a binomial in order to perform the following hypothesis test

$$H_0 : p = p_0, \tag{14.1}$$

verses

$$H_1 : p \neq p_0. \tag{14.2}$$

From the random sample we compute an estimate of p. In the binomial p represents the probability of a "success". Suppose we obtained x successes in n trials so $\widehat{p} = x/n$ is our point estimate of p. Then determine the statistic

$$z_0 = \frac{\widehat{p} - p_0}{\sqrt{p_0 q_0 / n}}, \tag{14.3}$$

where $q_0 = 1 - p_0$. If n is sufficiently large, say $n > 30$, then under the null hypothesis we may use the normal approximation to the binomial and z_0 is approximately $N(0,1)$ (Section 8.1 of [1]).

Let γ, $0 < \gamma < 1$, be the significance level of the test. Our decision rule is: (1) reject H_0 if $z_0 \geq z_{\gamma/2}$ or $z_0 \leq -z_{\gamma/2}$; and (2) do not reject H_0 when $-z_{\gamma/2} < z_0 < z_{\gamma/2}$.

14.3 Fuzzy Test

In Chapter 5 our fuzzy estimator for p is a triangular shaped fuzzy number $\overline{p}$ where its α-cuts are

$$\overline{p}[\alpha] = [\widehat{p} - z_{\alpha/2}\sqrt{\widehat{p}\widehat{q}/n}, \widehat{p} + z_{\alpha/2}\sqrt{\widehat{p}\widehat{q}/n}], \tag{14.4}$$

where $\widehat{q} = 1 - \widehat{p}$.

Calculations will be performed by alpha-cuts and interval arithmetic. Substitute $\overline{p}$ for $\widehat{p}$ in equation 14.3 and we obtain the following alpha-cut of our fuzzy statistic $\overline{Z}$

$$\overline{Z}[\alpha] = [z_0 - z_{\alpha/2}\sqrt{\frac{\widehat{p}\widehat{q}}{p_0 q_0}}, z_0 + z_{\alpha/2}\sqrt{\frac{\widehat{p}\widehat{q}}{p_0 q_0}}]. \tag{14.5}$$

The critical region will now be determined by fuzzy critical values $\overline{CV}_i$, $i = 1, 2$. They are determined as in the previous two chapters and they are given by their alpha-cuts

$$\overline{CV}_2[\alpha] = [z_{\gamma/2} - z_{\alpha/2}\sqrt{\frac{\widehat{p}\widehat{q}}{p_0 q_0}}, z_{\gamma/2} + z_{\alpha/2}\sqrt{\frac{\widehat{p}\widehat{q}}{p_0 q_0}}], \tag{14.6}$$

all α where γ is fixed, and because $\overline{CV}_1 = -\overline{CV}_2$

$$\overline{CV}_1[\alpha] = [-z_{\gamma/2} - z_{\alpha/2}\sqrt{\frac{\widehat{p}\widehat{q}}{p_0 q_0}}, -z_{\gamma/2} + z_{\alpha/2}\sqrt{\frac{\widehat{p}\widehat{q}}{p_0 q_0}}]. \tag{14.7}$$

Now that we have $\overline{Z}$, $\overline{CV}_1$ and $\overline{CV}_2$ we may compare $\overline{Z}$ and $\overline{CV}_1$ and then compare $\overline{Z}$ with $\overline{CV}_2$. The final decision rule was presented in Chapter 12.

Example 14.3.1

Let $n = 100$, $p_0 = q_0 = 0.5$, $\widehat{p} = 0.54$ so $\widehat{q} = 0.46$, and $\gamma = 0.05$. We then calculate $z_{\gamma/2} = 1.96$, and $z_0 = 0.80$. This is enough information to construct the fuzzy numbers $\overline{Z}$, $\overline{CV}_1$ and $\overline{CV}_2$ from Maple [2].

We first compare $\overline{Z}$ and $\overline{CV}_2$ to see which $\overline{Z} < \overline{CV}_2$, $\overline{Z} \approx \overline{CV}_2$ or $\overline{Z} > \overline{CV}_2$ is true. Since $z_0 < z_{\gamma/2}$ all we need to do is compare the right side of $\overline{Z}$ to the left side of $\overline{CV}_2$. From Figure 14.1 we see that the height of the intersection is less than 0.8 so we conclude that $\overline{Z} < \overline{CV}_2$. The Maple commands for Figure 14.1 are in Chapter 29.

Next we compare $\overline{Z}$ and $\overline{CV}_1$. Since $z_0 > -z_{\gamma/2}$ we compare the left side of $\overline{Z}$ to the right side of $\overline{CV}_1$. We easily see that $\overline{CV}_1 < \overline{Z}$.

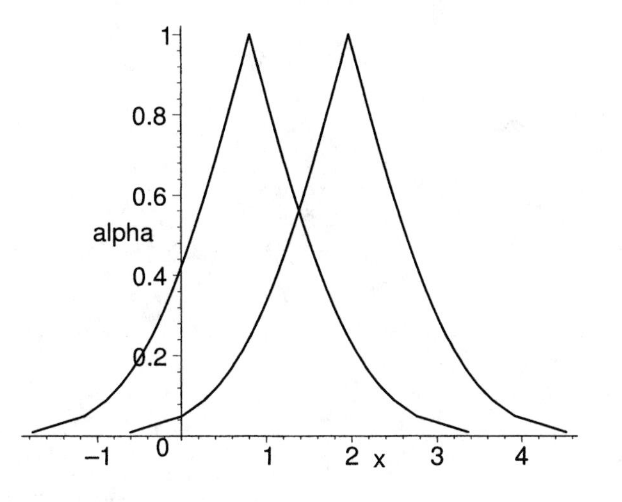

Figure 14.1: Fuzzy Test $\overline{Z}$ verses $\overline{CV}_2$ in Example 14.3.1($\overline{Z}$ left, $\overline{CV}_2$ right)

We do not reject H_0 because $\overline{CV}_1 < \overline{Z} < \overline{CV}_2$, which is the same result as in the crisp test.

We note again, as in previous chapters, that the graph of $\overline{Z}$ in Figure 14.1 is not accurate to the left of the vertical axis. This will also effect the left side of $\overline{CV}_2$. However, this does not effect the conclusion.

Example 14.3.2

Assume all the data is the same as in Example 14.3.1 except $\widehat{p} = 0.65$ so $\widehat{q} = 0.35$. We compute $z_0 = 3.00$. Now $z_0 > z_{\gamma/2} = 1.96$ so let us start with comparing the left side of $\overline{Z}$ to the right side of $\overline{CV}_2$. This is shown in Figure 14.2. We decide that $\overline{CV}_2 < \overline{Z}$ because the height of the intersection in Figure 14.2 is less than 0.8. Clearly we also obtain $\overline{CV}_1 < \overline{Z}$. Hence, we reject H_0.

14.4 References

1. R.V.Hogg and E.A.Tanis: Probability and Statistical Inference, Sixth Edition, Prentice Hall, Upper Saddle River, N.J., 2001.

2. Maple 6, Waterloo Maple Inc., Waterloo, Canada.

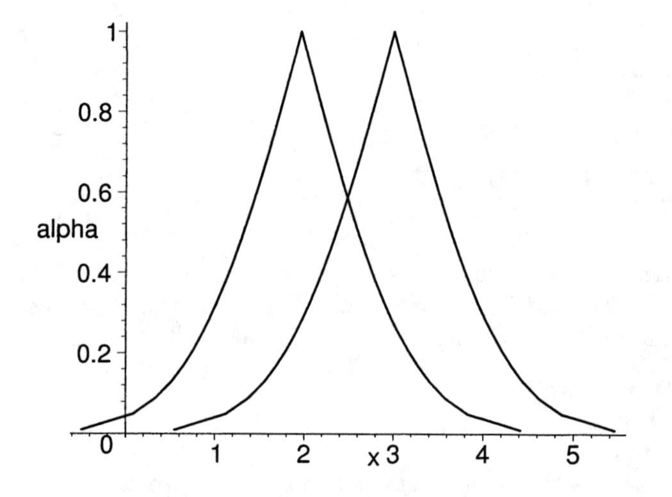

Figure 14.2: Fuzzy Test $\overline{Z}$ verses $\overline{CV}_2$ in Example 14.3.2($\overline{Z}$ right, $\overline{CV}_2$ left)

Chapter 15

Tests on σ^2, Normal Population

15.1 Introduction

We first describe the crisp hypothesis test and then construct the fuzzy test.

15.2 Crisp Hypothesis Test

We obtain a random sample of size n from a $N(\mu, \sigma^2)$, variance σ^2 unknown, in order to do the following hypothesis test

$$H_0 : \sigma^2 = \sigma_0^2, \tag{15.1}$$

verses

$$H_1 : \sigma^2 \neq \sigma_0^2. \tag{15.2}$$

From the random sample we compute its sample variance s^2 and then determine the statistic

$$\chi_0^2 = \frac{(n-1)s^2}{\sigma_0^2}. \tag{15.3}$$

Let γ, $0 < \gamma < 1$, be the significance level of the test. Now under the null hypothesis H_0 χ_0^2 has a chi-square distribution with $(n-1)$ degrees of freedom (Section 8.2 in [1]). Our decision rule is: (1) reject H_0 if $\chi_0^2 \geq \chi^2_{R,\gamma/2}$ or $\chi^2 \leq \chi^2_{L,\gamma/2}$; and (2) do not reject H_0 when $\chi^2_{L,\gamma/2} < \chi_0^2 < \chi^2_{R,\gamma/2}$. In the above decision rule $\chi^2_{L,\gamma/2}$ ($\chi^2_{R,\gamma/2}$) is the χ^2-value, on the left (right) side of the density, so that the probability of less than (exceeding) it is $\gamma/2$.

15.3 Fuzzy Hypothesis Test

Our fuzzy estimator of σ^2 is the triangular shaped fuzzy number $\overline{\sigma}^2$ having α-cuts given by equation (6.10) in Chapter 6. Now substitute $\overline{\sigma}^2$ in for s^2 in equation (15.3) and simplify using interval arithmetic and we obtain our fuzzy statistic $\overline{\chi}^2$ whose alpha-cuts are given by

$$\overline{\chi}^2[\alpha] = [\frac{n-1}{L(\lambda)}\chi_0^2, \frac{n-1}{R(\lambda)}\chi_0^2], \tag{15.4}$$

where $L(\lambda)$ ($R(\lambda)$) was defined in equation (6.8) ((6.9)) in Chapter 6. Recall that α is determined from λ, $0 \leq \lambda \leq 1$, as shown in equations (6.12) and (6.13) in Chapter 6.

Using this fuzzy statistic we determine the two fuzzy critical value sets whose alpha-cuts are

$$\overline{CV}_1[\alpha] = [\frac{n-1}{L(\lambda)}\chi_{L,\gamma/2}^2, \frac{n-1}{R(\lambda)}\chi_{L,\gamma/2}^2], \tag{15.5}$$

and

$$\overline{CV}_2[\alpha] = [\frac{n-1}{L(\lambda)}\chi_{R,\gamma/2}^2, \frac{n-1}{R(\lambda)}\chi_{R,\gamma/2}^2]. \tag{15.6}$$

In the above equations for the $\overline{CV}_i$ γ is fixed and λ ranges from zero to one. Notice that now $\overline{CV}_1 \neq -\overline{CV}_2$. When using the normal, or the t, distribution we were able to use $\overline{CV}_1 = -\overline{CV}_2$ because those densities were symmetric with respect to zero.

Having constructed these fuzzy numbers we go on to deciding on $\overline{\chi}^2 < \overline{CV}_1$,.....,$\overline{\chi}^2 > \overline{CV}_2$ and then our method of making the final decision (reject, do not reject, no decision) was outlined in Chapter 12.

Example 15.3.1

Let $n = 101$, $\sigma_0^2 = 2$, $\gamma = 0.01$ and from the random sample $s^2 = 1.675$. Then compute $\chi_0^2 = 83.75$, $\chi_{L,0.005}^2 = 67.328$, $\chi_{R,0.005}^2 = 140.169$.

Figure 15.1 shows $\overline{CV}_1$ on the left, $\overline{\chi}^2$ in the middle and $\overline{CV}_2$ on the right. The height of the intersection between $\overline{CV}_1$ and $\overline{\chi}^2$, and between $\overline{\chi}^2$ and $\overline{CV}_2$, are both below 0.8. We conclude $\overline{CV}_1 < \overline{\chi}^2 < \overline{CV}_2$ and we do not reject H_0. The Maple [2] commands for Figure 15.1 are in Chapter 29. The crisp test also concludes do not reject H_0.

Example 15.3.2

Assume everything is the same as in Example 15.3.1 except that $s^2 = 2.675$, Then $\chi_0^2 = 133.75$. The graphs of all three fuzzy numbers are shown in Figure 15.2. In Figure 15.2 $\overline{\chi}^2$ and $\overline{CV}_2$ are very close together, with $\overline{\chi}^2$ just to the

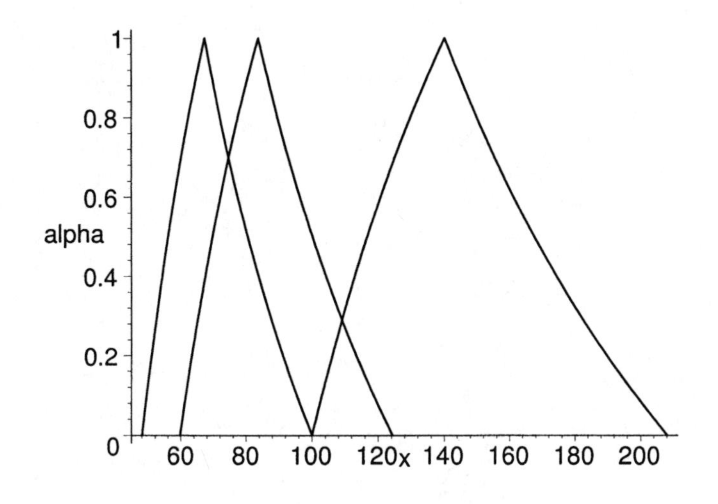

Figure 15.1: Fuzzy Test in Example 15.3.1($\overline{CV}_1$ left, $\overline{\chi}^2$ middle, $\overline{CV}_2$ right)

left of $\overline{CV}_2$, so $\overline{CV}_2 \approx \overline{\chi}^2$. The final result is $\overline{CV}_1 < \overline{\chi}^2 \approx \overline{CV}_2$ and we have no decision on H_0. The crisp test would decide to not reject H_0.

15.4 References

1. R.V.Hogg and E.A.Tanis: Probability and Statistical Inference, Sixth Edition, Prentice Hall, Upper Saddle River, N.J., 2001.

2. Maple 6, Waterloo Maple Inc., Waterloo, Canada.

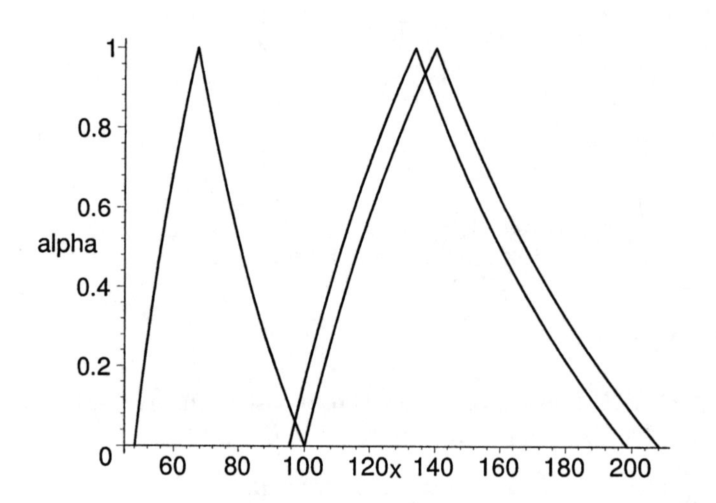

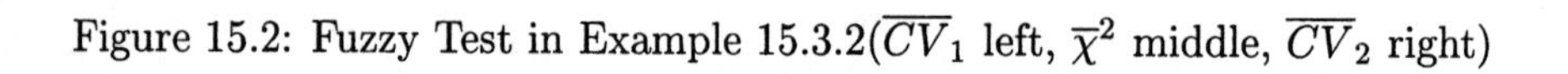
Figure 15.2: Fuzzy Test in Example 15.3.2($\overline{CV}_1$ left, $\overline{\chi}^2$ middle, $\overline{CV}_2$ right)

Chapter 16

Tests μ_1 verses μ_2, Variances Known

16.1 Introduction

Let us first review the details on the crisp test.

16.2 Non-Fuzzy Test

As in Chapter 7 we have two populations: Pop I and Pop II. Pop I is normally distributed with unknown mean μ_1 and known variance σ_1^2. Pop II is also normally distributed with unknown mean μ_2 but known variance σ_2^2. We wish to perform the following test

$$H_0 : \mu_1 - \mu_2 = 0, \tag{16.1}$$

verses

$$H_1 : \mu_1 - \mu_2 \neq 0. \tag{16.2}$$

Or what is the same thing, we want to test $\mu_1 = \mu_2$ verses $\mu_1 \neq \mu_2$.

We collect a random sample of size n_1 from Pop I and let $\overline{x}_1$ be the mean for this data. We also gather a random sample of size n_2 from Pop II and $\overline{x}_2$ is the mean for the second sample. We assume these two random samples are independent.

Now, under the null hypothesis, $\overline{x}_1 - \overline{x}_2$ is normally distributed with mean $\mu_1 - \mu_2$ and standard deviation $\sigma_0 = \sqrt{\sigma_1^2/n_1 + \sigma_2^2/n_2}$ (Section 8.3 of [1]). Our test statistic is

$$z_0 = \frac{(\overline{x}_1 - \overline{x}_2) - 0}{\sigma_0}. \tag{16.3}$$

Let γ, $0 < \gamma < 1$, be the significance level of the test. Usual values for γ are $0.10, 0.05, 0.01$. Now under the null hypothesis H_0 z_0 is $N(0,1)$ and our decision rule is: (1) reject H_0 if $z_0 \geq z_{\gamma/2}$ or $z_0 \leq -z_{\gamma/2}$; and (2) do not reject H_0 when $-z_{\gamma/2} < z_0 < z_{\gamma/2}$. The numbers $\pm z_{\gamma/2}$ are called the critical values (cv's) for the test. In the above decision rule $z_{\gamma/2}$ is the z-value so that the probability of a random variable, having the $N(0,1)$ probability density, exceeding z is $\gamma/2$.

16.3 Fuzzy Test

Now proceed to the fuzzy situation where our fuzzy estimate of $\mu_1 - \mu_2$, as explained in Chapter 7, is the triangular shaped fuzzy number $\overline{\mu}_{12}$ and its α-cuts are given in equation (7.1) in Chapter 7. Substitute $\overline{\mu}_{12}$ in for $\overline{x}_1 - \overline{x}_2$ in equation (16.3), simplify using interval arithmetic, producing our fuzzy statistic $\overline{Z}$ whose α-cuts are

$$\overline{Z}[\alpha] = [z_0 - z_{\alpha/2}, z_0 + z_{\alpha/2}]. \tag{16.4}$$

We put these α-cuts together to get a fuzzy triangular shaped fuzzy number $\overline{Z}$.

The critical region will now be determined by fuzzy critical values $\overline{CV}_i$, $i = 1, 2$. They are determined as in the previous chapters and they are given by their alpha-cuts

$$\overline{CV}_2[\alpha] = [z_{\gamma/2} - z_{\alpha/2}, z_{\gamma/2} + z_{\alpha/2}], \tag{16.5}$$

all α where γ is fixed, and since $\overline{CV}_1 = -\overline{CV}_2$

$$\overline{CV}_1[\alpha] = [-z_{\gamma/2} - z_{\alpha/2}, -z_{\gamma/2} + z_{\alpha/2}]. \tag{16.6}$$

Now that we have $\overline{Z}$, $\overline{CV}_1$ and $\overline{CV}_2$ we may compare $\overline{Z}$ and $\overline{CV}_1$ and then compare $\overline{Z}$ with $\overline{CV}_2$. The final decision rule was presented in Chapter 12.

Example 16.3.1

Assume that: (1) $n_1 = 15$, $\overline{x}_1 = 70.1$,$\sigma_1^2 = 6$; and (2) $n_2 = 8$, $\overline{x}_2 = 75.3$,$\sigma_2^2 = 4$. Also let $\gamma = 0.05$ so that $z_{\gamma/2} = 1.96$. We compute $\sigma_0 = 0.9487$ and $z_0 = -5.48$. Since $z_0 < -z_{\gamma/2}$ we compare $\overline{Z}$ and only $\overline{CV}_1$. This is shown in Figure 16.1. Clearly $\overline{Z} < \overline{CV}_1$ (because the height of the intersection is less than 0.8) and $\overline{Z} < \overline{CV}_2$ so we reject H_0. The result is the same for the crisp test. The Maple [2] commands for Figure 16.1 are in Chapter 29.

16.4 References

1. R.V.Hogg and E.A.Tanis: Probability and Statistical Inference, Sixth Edition, Prentice Hall, Upper Saddle River, N.J., 2001.

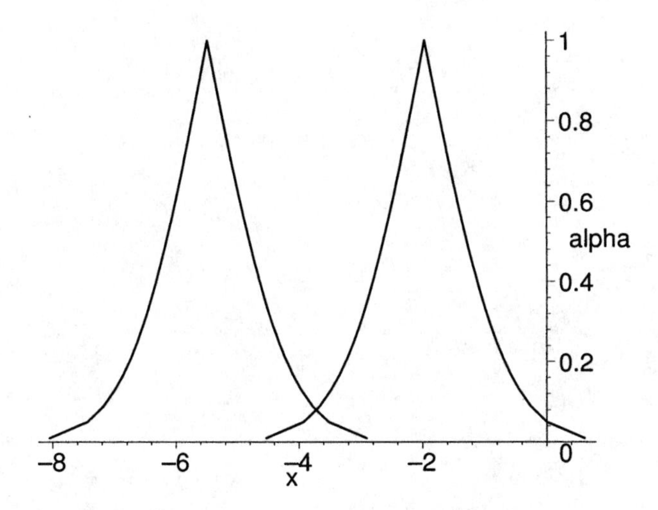

Figure 16.1: Fuzzy Test $\overline{Z}$ verses $\overline{CV}_1$ in Example 16.3.1($\overline{Z}$ left, $\overline{CV}_1$ right)

2. Maple 6, Waterloo Maple Inc., Waterloo, Canada

Chapter 17

Test μ_1 verses μ_2, Variances Unknown

17.1 Introduction

We have two populations: Pop I and Pop II. Pop I is normally distributed with unknown mean μ_1 and unknown variance σ_1^2. Pop II is also normally distributed with unknown mean μ_2 and unknown variance σ_2^2. We wish to do the following statistical test

$$H_0 : \mu_1 - \mu_1 = 0, \tag{17.1}$$

verses

$$H_1 : \mu_1 - \mu_1 \neq 0. \tag{17.2}$$

We collect a random sample of size n_1 from Pop I and let $\overline{x}_1$ be the mean for this data and s_1^2 is the sample variance. We also gather a random sample of size n_2 from Pop II and $\overline{x}_2$ is the mean for the second sample with s_2^2 the variance. We assume these two random samples are independent.

There will be two cases: (1) large samples; and (2) small samples. Within the "small samples" we will consider two cases: (1) the population variances are (approximately) equal; and (2) the population variances are not equal. In all cases we first review the crisp test first.

17.2 Large Samples

By large samples we will mean that $n_1 > 30$ and $n_2 > 30$. The statistic for this test is

$$z_0 = \frac{(\overline{x}_1 - \overline{x}_2) - 0}{s_0}, \tag{17.3}$$

where

$$s_0 = \sqrt{\frac{s_1^2}{n_1} + \frac{s_2^2}{n_2}}. \tag{17.4}$$

Under H_0 z_0 is approximately normally distributed with zero mean and unit variance (Section 8.3 of [1]). Let γ, $0 < \gamma < 1$, be the significance level of the test. Our decision rule is: (1) reject H_0 if $z_0 \geq z_{\gamma/2}$ or $z_0 \leq -z_{\gamma/2}$; and (2) do not reject H_0 when $-z_{\gamma/2} < z_0 < z_{\gamma/2}$.

We may now develop the fuzzy test. In equation (17.3) we substitute: (1) $\overline{\mu}_{12}$, our fuzzy estimator for $\mu_1 - \mu_2$ in equation (8.1) of Chapter 8, for $\overline{x}_1 - \overline{x}_2$; (2) substitute $\overline{\sigma}_1^2$, our fuzzy estimator for σ_1^2 in equation (6.10) of Chapter 6, for s_1^2; and (3)substitute $\overline{\sigma}_2^2$, our fuzzy estimator for σ_2^2 in equation (6.10) of Chapter 6, for s_2^2. We then use alpha-cuts and interval arithmetic to simplify and we find that the α-cuts of the test statistic $\overline{Z}$ are

$$\overline{Z}[\alpha] = \frac{[a,b]}{\sqrt{[c_1,d_1] + [c_2,d_2]}}, \tag{17.5}$$

where: (1) $a = (\overline{x}_1 - \overline{x}_2) - z_{\alpha/2}s_0$; (2) $b = (\overline{x}_1 - \overline{x}_2) + z_{\alpha/2}s_0$; (3) $c_1 = [(n_1-1)s_1^2]/[L_1(\lambda)n_1]$; (4) $d_1 = [(n_1-1)s_1^2]/[R_1(\lambda)n_1]$; (5) $c_2 = [(n_2 - 1)s_2^2]/[L_2(\lambda)n_2]$; and (6) $d_2 = [(n_2-1)s_2^2]/[R_2(\lambda)n_2]$. And

$$L_1(\lambda) = [1-\lambda]\chi^2_{R,0.005} + \lambda(n_1 - 1), \tag{17.6}$$

$$R_1(\lambda) = [1-\lambda]\chi^2_{L,0.005} + \lambda(n_1 - 1), \tag{17.7}$$

where the degrees of freedom in the χ^2 distribution is $n_1 - 1$. Also

$$L_2(\lambda) = [1-\lambda]\chi^2_{R,0.005} + \lambda(n_2 - 1), \tag{17.8}$$

$$R_2(\lambda) = [1-\lambda]\chi^2_{L,0.005} + \lambda(n_2 - 1), \tag{17.9}$$

and the degrees of freedom for these χ^2 distributions is $n_2 - 1$. The $L_i(\lambda)$ ($R_i(\lambda)$) here is similar to the one defined in equation (6.8) ((6.9)) in Chapter 6. In the above equations for the $L_i(\lambda)$ and the $R_i(\lambda)$ we have $0 \leq \lambda \leq 1$. Then

$$\overline{Z}[\alpha] = \frac{[a,b]}{\sqrt{[c_1+c_2, d_1+d_2]}} = \tag{17.10}$$

$$= \frac{[a,b]}{[\sqrt{c_1+c_2}, \sqrt{d_1+d_2}]} = \tag{17.11}$$

$$= [\frac{a}{\sqrt{d_1+d_2}}, \frac{b}{\sqrt{c_1+c_2}}]. \tag{17.12}$$

In the last step we assumed that $a > 0$. If $a < 0$ and $b > 0$, or $b < 0$, we need to make some changes in the interval arithmetic.

Now substitute in for the values of $a,b,\ldots,d_2$ and we obtain

$$\overline{Z}[\alpha] = [\Gamma_R(\lambda)(z_0 - z_{\alpha/2}), \Gamma_L(\lambda)(z_0 + z_{\alpha/2})], \tag{17.13}$$

where

$$\Gamma_R(\lambda) = \frac{s_0}{s_R(\lambda)}, \tag{17.14}$$

$$\Gamma_L(\lambda) = \frac{s_0}{s_L(\lambda)}, \tag{17.15}$$

for

$$s_R(\lambda) = \sqrt{d_1 + d_2}, \tag{17.16}$$

$$s_L(\lambda) = \sqrt{c_1 + c_2}. \tag{17.17}$$

The above equations define out fuzzy test statistic $\overline{Z}$. As in Chapter 6, equations (6.12) and (6.13), we have $\alpha = f(\lambda)$ given by

$$\alpha_i = \int_0^{R_i(\lambda)} \chi^2 dx + \int_{L_i(\lambda)}^{\infty} \chi^2 dx, \tag{17.18}$$

for $i = 1, 2$. The χ^2 distribution has $n_1 - 1$ ($n_2 - 1$) degrees of freedom when $i = 1$ ($i = 2$). However, if $n_1 \neq n_2$ and $0 < \lambda < 1$ we may not get $\alpha_1 = \alpha_2$. We ran a number of numerical examples using Maple [2] for various values of $n_1 > 30$, $n_2 > 30$ and $n_1 \neq n_2$, and we found that $\alpha_1 \neq \alpha_2$. So for $n_1 \neq n_2$ we are unable to relate $\lambda \in [0, 1]$ to a single value for $\alpha \in [0.01, 1]$ for the alpha-cuts for $\overline{Z}$. In order to proceed let us now assume that $n_1 = n_2 = n$. An alternate procedure for "solving" the problem of $n_1 \neq n_2$ is at the end of this chapter.

With $n_1 = n_2 = n$ we obtain $L_1(\lambda) = L_2(\lambda) = L(\lambda)$, $R_1(\lambda) = R_2(\lambda) = R(\lambda)$, $s_L(\lambda) = \sqrt{(n-1)/L(\lambda)}s_0$, $s_R(\lambda) = \sqrt{(n-1)/R(\lambda)}s_0$, $\Gamma_R(\lambda) = \sqrt{R(\lambda)/(n-1)}$ and $\Gamma_L(\lambda) = \sqrt{L(\lambda)/(n-1)}$. Hence

$$\overline{Z}[\alpha] = [\Pi_1(z_0 - z_{\alpha/2}), \Pi_2(z_0 + z_{\alpha/2})]. \tag{17.19}$$

where

$$\Pi_1 = \sqrt{\frac{R(\lambda)}{n-1}}, \tag{17.20}$$

and

$$\Pi_2 = \sqrt{\frac{L(\lambda)}{n-1}}, \tag{17.21}$$

with $L(\lambda)$ ($R(\lambda)$) defined in equation (6.8) ((6.9)) in Chapter 6.

Having $n_1 = n_2 = n$ simplifies the calculations for $\overline{Z}[\alpha]$ as follows

$$\overline{Z}[\alpha] = [a, b]/\sqrt{[c, d]} = [a, b]/[\sqrt{c}, \sqrt{d}] = [a, b][\frac{1}{\sqrt{d}}, \frac{1}{\sqrt{c}}], \tag{17.22}$$

because $c > 0$, where a and b were given above, and now

$$c = \frac{n-1}{L(\lambda)} s_0^2, \tag{17.23}$$

and

$$d = \frac{n-1}{R(\lambda)} s_0^2. \tag{17.24}$$

To complete the calculation it depends if $b > a > 0$ or $a < b < 0$ for all alpha , or if for certain values of α we have $a < 0 < b$. This was all discussed in Subsection 13.3.1 in Chapter 13. In equation (17.19) we assumed that $0 < a < b$ for all $\alpha \in [0, 1]$. Then we got

$$\overline{Z}[\alpha] = [\frac{a}{\sqrt{d}}, \frac{b}{\sqrt{c}}]. \tag{17.25}$$

However, if $b < 0$ for certain values of alpha, as in the following example, we have

$$\overline{Z}[\alpha] = [\frac{a}{\sqrt{c}}, \frac{b}{\sqrt{d}}]. \tag{17.26}$$

Next we compute the fuzzy critical values as in previous chapters and we obtain (assuming $0 < a < b$)

$$\overline{CV}_2[\alpha] = [\Pi_1(z_{\gamma/2} - z_{\alpha/2}), \Pi_2(z_{\gamma/2} + z_{\alpha/2})], \tag{17.27}$$

and because $\overline{CV}_1 = -\overline{CV}_2$

$$\overline{CV}_1[\alpha] = [\Pi_2(-z_{\gamma/2} - z_{\alpha/2}), \Pi_1(-z_{\gamma/2} + z_{\alpha/2})]. \tag{17.28}$$

In these last two equations γ is fixed but alpha varies from 0.01 to one.

Example 17.2.1

Assume that: (1) $n_1 = 41$, $\overline{x}_1 = 6.701$,$s_1^2 = 0.108$ all from Pop I; and (2) from Pop II $n_2 = 41$, $\overline{x}_2 = 6.841$,$s_2^2 = 0.155$. Also let $\gamma = 0.05$ so that $z_{\gamma/2} = 1.96$. We compute $z_0 = -1.748$ and $s_0 = 0.0801$. We will also need $\chi^2_{L,0.005} = 20.707$ and $\chi^2_{R,0.005} = 66.766$, both for 40 degrees of freedom. Then $L(\lambda) = 66.766 - 26.766\lambda$ and $R(\lambda) = 20.707 + 19.293\lambda$. From this we may determine

$$\Pi_1 = \sqrt{0.517675 + 0.482325\lambda}, \tag{17.29}$$

and

$$\Pi_2 = \sqrt{1.66915 - 0.66915\lambda}. \tag{17.30}$$

Since $z_0 < 0$ we start with comparing $\overline{Z}$ and $\overline{CV}_1$. Using the notation defined above we have $a < 0$ all alpha but $b < 0$ for $0 < \alpha^* < \alpha \leq 1$. So, for $\overline{Z}[\alpha]$ and $0 < \alpha^* < \alpha \leq 1$ we use $[\frac{a}{\sqrt{c}}, \frac{b}{\sqrt{d}}]$ which produces

$$\overline{Z}[\alpha] = [\Pi_2(z_0 - z_{\alpha/2}), \Pi_1(z_0 + z_{\alpha/2})]. \tag{17.31}$$

Since $\overline{Z}$ changed the $\overline{CV}_i$ will change

$$\overline{CV}_1[\alpha] = [\Pi_2(-z_{\gamma/2} - z\alpha/2), \Pi_1(-z_{\gamma/2} + z_{\alpha/2})], \tag{17.32}$$

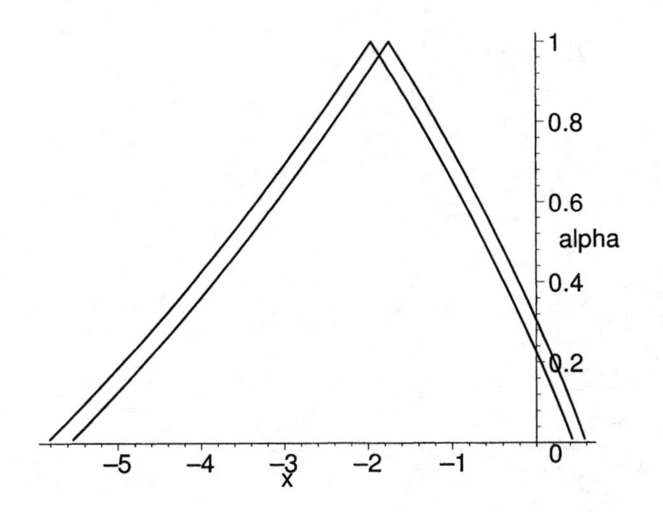

Figure 17.1: Fuzzy Test $\overline{Z}$ verses $\overline{CV}_1$ in Example 17.2.1($\overline{Z}$ right,$\overline{CV}_1$ left)

and since $\overline{CV}_2 = -\overline{CV}_1$

$$\overline{CV}_2[\alpha] = [\Pi_1(z_{\gamma/2} - z\alpha/2), \Pi_2(z_{\gamma/2} + z_{\alpha/2})], \tag{17.33}$$

where γ is fixed but alpha varies from α^* to one. For $0.01 \leq \alpha < \alpha^*$ we switch and use $\overline{Z}[\alpha] = [\frac{a}{\sqrt{c}}, \frac{b}{\sqrt{c}}]$. However, in this case this switch will not effect the results.

The result of $\overline{Z}$ verse $\overline{CV}_1$ is shown in Figure 17.1. Clearly $\overline{Z} \approx \overline{CV}_1$ (because the height of the intersection is more than 0.8) and $\overline{Z} < \overline{CV}_2$ so we make no decision on H_0. The result in the crisp case would be to not reject H_0. The Maple [2] commands for Figure 17.1 are in Chapter 29.

We should point out that the graphs in Figure 17.1 are not completely correct. They are correct to the left of the vertical axis ($a < b < 0$), but they are not correct to the right of the vertical axis ($a < 0 < b$) because we did not make that "switch" mentioned above. This does not change the final result

Example 17.2.2

Assume that: (1) $n_1 = 61$, $\overline{x}_1 = 53.3$,$s_1^2 = 12.96$ all from Pop I; and (2) from Pop II $n_2 = 61$, $\overline{x}_2 = 51.3$,$s_2^2 = 20.25$. Also let $\gamma = 0.05$ so that $z_{\gamma/2} = 1.96$. We compute $z_0 = 2.71$ and $s_0 = 0.7378$. We will also need $\chi^2_{L,0.005} = 35.534$ and $\chi^2_{R,0.005} = 91.952$ both for degrees of freedom 60. Then $L(\lambda) = 91.952 - 31.952\lambda$ and $R(\lambda) = 35.534 + 24.466\lambda$. From this we

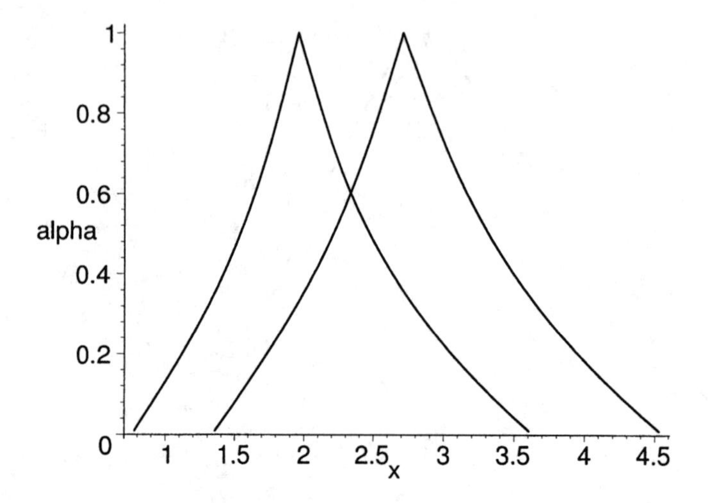

Figure 17.2: Fuzzy Test $\overline{Z}$ verses $\overline{CV}_2$ in Example 17.2.2($\overline{Z}$ right,$\overline{CV}_2$ left)

may determine

$$\Pi_1 = \sqrt{0.5922 + 0.4078\lambda}, \tag{17.34}$$

and

$$\Pi_2 = \sqrt{1.5325 - 0.5325\lambda}. \tag{17.35}$$

Since $z_0 > 0$ we start with comparing $\overline{Z}$ and $\overline{CV}_2$. Because $0 < a$ for $0 \leq \alpha \leq 1$, see Figure 17.2, we may use equation (17.19) for $\overline{Z}[\alpha]$ and we use equation (17.27) for $\overline{CV}_2$ for the comparison.

The comparison of $\overline{Z}$ and $\overline{CV}_2$ is shown in Figure 17.2. We see that $\overline{CV}_2 < \overline{Z}$, because the height of the intersection is less than 0.8, we reject H_0. The crisp test would also reject the null hypothesis.

17.3 Small Samples

By small samples we mean $n \leq 30$. We have assumed, as discussed in the previous section, that both sample sizes are equal to n. There will be two cases to consider: (1) the first case is where we may assume that $\sigma_1^2 = \sigma_2^2 = \sigma^2$; and (2) the other is where the variances are different.

17.3.1 Equal Variances

The test statistic is (Section 8.3 in [1])

$$t_0 = \frac{(\overline{x}_1 - \overline{x}_2) - 0}{s_0}, \tag{17.36}$$

where s_0 was defined in equation (17.4). Actually, the test statistic is defined differently but it simplifies to this if $n_1 = n_2 = n$. In the original test statistic the denominator in the above equation is

$$(s_p)\sqrt{\frac{1}{n_1} + \frac{1}{n_2}}, \tag{17.37}$$

where

$$s_p = \sqrt{\frac{(n_1 - 1)s_1^2 + (n_2 - 1)s_2^2}{n_1 + n_2 - 2}}. \tag{17.38}$$

Now let $n_1 = n_2 = n$ and we get s_0 in the denominator. Under H_0 t_0 has a t-distribution with $2(n-1)$ degrees of freedom.

This case is similar to that studied in Section 17.2 except substitute the t-distribution for the normal distribution. Therefore, we will not pursue this case any further at this time.

17.3.2 Unequal Variances

The test statistic is (Section 8.3 in [1])

$$t_0 = \frac{(\overline{x}_1 - \overline{x}_2) - 0}{s_0}. \tag{17.39}$$

Assuming equal sample sizes this test statistic is the same as in Section 17.2 except now we have the t-distribution with r degrees of freedom and r is calculated as in Section 8.3.2 in Chapter 8.

Following the development of Section 17.2, except now substitute our fuzzy estimator in equation (8.11) in Chapter 8 for $(\overline{x}_1 - \overline{x}_2)$, and assuming all intervals are positive, we obtain

$$\overline{T}[\alpha] = [\Pi_1(t_0 - t_{\alpha/2}), \Pi_2(t_0 + t_{\alpha/2})], \tag{17.40}$$

where the t-distribution has r degrees of freedom. Then

$$\overline{CV}_2[\alpha] = [\Pi_1(t_{\gamma/2} - t_{\alpha/2}), \Pi_2(t_{\gamma/2} + t_{\alpha/2})], \tag{17.41}$$

and because $\overline{CV}_1 = -\overline{CV}_2$

$$\overline{CV}_1[\alpha] = [\Pi_2(-t_{\gamma/2} - t_{\alpha/2}), \Pi_1(-t_{\gamma/2} + t_{\alpha/2})]. \tag{17.42}$$

In these last two equations γ is fixed but alpha varies from 0.01 to one.

Example 17.3.2.1

Let us use the same data as in Example 17.2.2, except for small samples $n_1 = n_2 = 21$. We have the same value for $s_1^2 = 12.96$, $s_2^2 = 20.25$, $\gamma = 0.05$

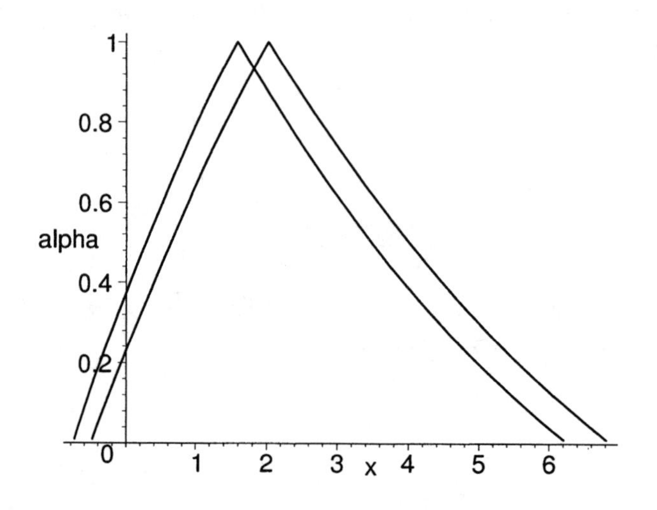

Figure 17.3: Fuzzy Test $\overline{T}$ verses $\overline{CV}_2$ in Example 17.3.2.1($\overline{T}$ left,$\overline{CV}_2$ right)

and then $t_0 = 1.5905$ with $s_0 = 1.2575$. We find the new values of the Π_i, using 20 degrees of freedom in the χ^2 distributions, giving

$$\Pi_1 = \sqrt{0.3717 + 0.6283\lambda}, \tag{17.43}$$

and

$$\Pi_2 = \sqrt{1.9998 - 0.9998\lambda}. \tag{17.44}$$

Lastly, we need to find the degrees of freedom r for the t-distribution. Using equation (8.8) of Chapter 8 we find the degrees of freedom is 39. Then $t_{\gamma/2} = 2.022$. Since $t_0 > 0$ we start with comparing $\overline{T}$ and $\overline{CV}_2$. We use equation (17.40) for $\overline{T}$ and equation (17.41) for $\overline{CV}_2$.

The comparison of $\overline{T}$ and $\overline{CV}_2$ is shown in Figure 17.3. We see that $\overline{CV}_2 \approx \overline{Z}$, because the height of the intersection is greater than 0.8. Clearly, $\overline{CV}_1 < \overline{T}$. Therefore, we make no decision on H_0. The crisp test would reject the null hypothesis. Maple [2] commands for Figure 17.3 are in Chapter 29.

The graph of $\overline{T}$ is not completely correct to the left of the vertical axis. This also effects the left side of the graph of $\overline{CV}_2$. However, we still obtain the same conclusion.

17.4 $n_1 \neq n_2$

We first discuss changes in Section 17.2 and then Section 17.3.

The α_i obtained from equation (17.18) give: (1) a $(1-\alpha_1)100\%$ confidence interval for σ_1^2; and (2) a $(1-\alpha_2)100\%$ confidence interval for σ_2^2. We need an α for the $(1-\alpha)100\%$ confidence interval for $\mu_1 - \mu_2$. As a compromise solution let

$$\alpha = \frac{\alpha_1 + \alpha_2}{2}. \tag{17.45}$$

Then, assuming all intervals are positive, equation (17.13) is $\overline{Z}[\alpha]$ and

$$\overline{CV}_2[\alpha] = [\Gamma_R(\lambda)(z_{\gamma/2} - z_{\alpha/2}), \Gamma_L(\lambda)(z_{\gamma/2} + z_{\alpha/2})], \tag{17.46}$$

with $\overline{CV}_1 = -\overline{CV}_2$. Then we may construct graphs of $\overline{Z}$ and the $\overline{CV}_i$.

Now let us discuss Section 17.3. We start with Section 17.3.1. The test statistic is

$$t_0 = \frac{\overline{x}_1 - \overline{x}_2}{s^*}, \tag{17.47}$$

with s^* defined in Section 8.3.1 of Chapter 8.

We may now develop the fuzzy test. In equation (17.47) we substitute: (1) $\overline{\mu}_{12}$, our fuzzy estimator for $\mu_1 - \mu_2$ in equation (8.5) of Chapter 8, for $\overline{x}_1 - \overline{x}_2$; (2) substitute $\overline{\sigma}_1^2$, our fuzzy estimator for σ_1^2 in equation (6.10) of Chapter 6, for s_1^2; and (3)substitute $\overline{\sigma}_2^2$, our fuzzy estimator for σ_2^2 in equation (6.10) of Chapter 6, for s_2^2. We then use alpha-cuts and interval arithmetic to simplify, assuming all intervals are positive, and we find that the α-cuts of the test statistic $\overline{T}$ are

$$[\Upsilon_R(\lambda)(t_0 - t_{\beta/2}), \Upsilon_L(\lambda)(t_0 + t_{\beta/2}], \tag{17.48}$$

where

$$\Upsilon_R(\lambda) = \frac{s_p}{s_R(\lambda)}, \tag{17.49}$$

$$\Upsilon_L(\lambda) = \frac{s_p}{s_L(\lambda)}, \tag{17.50}$$

$$s_R(\lambda) = \sqrt{s_{R1}(\lambda) + s_{R2}(\lambda)}, \tag{17.51}$$

$$s_L(\lambda) = \sqrt{s_{L1}(\lambda) + s_{L2}(\lambda)}, \tag{17.52}$$

$$s_{R1}(\lambda) = \frac{(n_1 - 1)^2 s_1^2}{R_1(\lambda)(n_1 + n_2 - 2)}, \tag{17.53}$$

$$s_{R2}(\lambda) = \frac{(n_2 - 1)^2 s_2^2}{R_2(\lambda)(n_1 + n_2 - 2)}, \tag{17.54}$$

$$s_{L1}(\lambda) = \frac{(n_1 - 1)^2 s_1^2}{L_1(\lambda)(n_1 + n_2 - 2)}, \tag{17.55}$$

$$s_{L2}(\lambda) = \frac{(n_2 - 1)^2 s_2^2}{L_2(\lambda)(n_1 + n_2 - 2)}. \tag{17.56}$$

The $L_i(\lambda)$ and $R_i(\lambda)$ were defined in Section 17.2.

Using $\alpha = (\alpha_1 + \alpha_2)/2$ we may now construct the graph of $\overline{T}$. Also

$$\overline{CV}_2[\alpha] = [\Upsilon_R(\lambda)(t_{\gamma/2} - t_{\alpha/2}), \Upsilon_L(\lambda)(t_{\gamma/2} + t_{\alpha/2})], \tag{17.57}$$

and $\overline{CV}_1 = -\overline{CV}_2$. Then the graphs, and comparisons, of $\overline{T}$ and the $\overline{CV}_i$ can be done.

Lastly we look at Section 17.3.2. This case is similar to Section 17.2 except now we have the t-distribution with r degrees of freedom and r is calculated as in Section 8.3.2 in Chapter 8. We shall not consider this case, for $n_1 \neq n_2$, any further.

17.5 References

1. R.V.Hogg and E.A.Tanis: Probability and Statistical Inference, Sixth Edition, Prentice hall, Upper Saddle River, N.J., 2001.

2. Maple 6, Waterloo Maple Inc., Waterloo, Canada

Chapter 18

Test $p_1 = p_2$, Binomial Populations

18.1 Non-Fuzzy Test

In this chapter we have two binomial populations: Pop I and Pop II. In Pop I (II) let p_1 (p_2) be the probability of a "success". We want a do the following test

$$H_0 : p_1 - p_2 = 0, \tag{18.1}$$

verses

$$H_1 : p_1 - p_2 \neq 0. \tag{18.2}$$

We take a random sample of size n_1 (n_2) from Pop I (II) and observe x_1 (x_2) successes. Then our point estimator for p_1 (p_2) is $\widehat{p}_1 = x_1/n_1$ ($\widehat{p}_2 = x_2/n_2$). We assume that these two random samples are independent. Let $\widehat{q}_i = 1 - \widehat{p}_i$, $i = 1, 2$.

We would like to use the normal approximation to the binomial. To do this we now assume that n_1 and n_2 are sufficiently large for the normal approximation. Under the null hypothesis let $p_1 = p_2 = p$ with $q = 1 - p$. So, assuming H_0 is true, $\widehat{p}_1$ ($\widehat{p}_2$) is approximately normally distributed with mean p and variance pq/n_1 (pq/n_2). It follows that $p_1 - p_2$ is approximately normally distributed with mean zero and variance $pq/n_1 + pq/n_2$. Then our test statistic

$$z_0 = [(\widehat{p}_1 - \widehat{p}_2) - 0]/[\sqrt{pq/n_1 + pq/n_2}], \tag{18.3}$$

is also approximately normally distributed with zero mean and unit variance.

Let γ, $0 < \gamma < 1$, be the significance level of the test. Our decision rule is : (1) reject H_0 if $z_0 \geq z_{\gamma/2}$ or $z_0 \leq -z_{\gamma/2}$; and (2) do not reject H_0 when $-z_{\gamma/2} < z_0 < z_{\gamma/2}$.

However, our test is not operational because we have no values to use for p and q in our test statistic. To make it operational we substitute an estimate

$\widehat{p}$ of p ($\widehat{q} = 1 - \widehat{p}$) into equation (18.3). The estimator usually used is

$$\widehat{p} = \frac{x_1 + x_2}{n_1 + n_2}, \tag{18.4}$$

or the "pooled" estimator of p since under H_0 $p_1 = p_2 = p$. The final test statistic (Section 8.1 in [1]), which can be used to compute a value in the hypothesis test, is

$$z_0 = [(\widehat{p}_1 - \widehat{p}_2) - 0]/\sqrt{\widehat{pq}/n_1 + \widehat{pq}/n_2}. \tag{18.5}$$

18.2 Fuzzy Test

We substitute our fuzzy estimator $\overline{p}_{12}$, whose α-cuts are given in equation (10.3) in Chapter 10, for $\widehat{p}_1 - \widehat{p}_2$, in equation (18.5), to get our fuzzy statistic $\overline{Z}$. After simplification we obtain

$$\overline{Z}[\alpha] = [z_0 - z_{\alpha/2}\sqrt{A/B}, z_0 + z_{\alpha/2}\sqrt{A/B}], \tag{18.6}$$

where

$$A = \frac{\widehat{p}_1\widehat{q}_1}{n_1} + \frac{\widehat{p}_2\widehat{q}_2}{n_2}, \tag{18.7}$$

and

$$B = \frac{\widehat{pq}}{n_1} + \frac{\widehat{pq}}{n_2}. \tag{18.8}$$

The critical region will now be determined by fuzzy critical values $\overline{CV}_i$, $i = 1, 2$. They are found as in the previous chapters and they are given by their alpha-cuts

$$\overline{CV}_2[\alpha] = [z_{\gamma/2} - z_{\alpha/2}\sqrt{A/B}, z_{\gamma/2} + z_{\alpha/2}\sqrt{A/B}], \tag{18.9}$$

all α where γ is fixed, and because $\overline{CV}_1 = -\overline{CV}_2$

$$\overline{CV}_1[\alpha] = [-z_{\gamma/2} - z_{\alpha/2}\sqrt{A/B}, -z_{\gamma/2} + z_{\alpha/2}\sqrt{A/B}]. \tag{18.10}$$

Now that we have $\overline{Z}$, $\overline{CV}_1$ and $\overline{CV}_2$ we may compare $\overline{Z}$ and $\overline{CV}_1$ and then compare $\overline{Z}$ with $\overline{CV}_2$. The final decision rule was presented in Chapter 12.

Example 18.2.1

Let the data be: (1) $n_1 = 200$, $x_1 = 30$; and (2) $n_2 = 1000$, $x_2 = 200$. From this we compute $\widehat{p}_1 = 0.15$, $\widehat{q}_1 = 0.85$, $\widehat{p}_2 = 0.20$, $\widehat{q}_2 = 0.80$, $\widehat{p} = 0.19$, and $\widehat{q} = 0.89$. Using these values we determine $z_0 = -1.645$, and $\sqrt{A/B} = 0.9293$.

Let $\gamma = 0.05$ so that $z_{\gamma/2} = 1.96$. This is sufficient to get the graphs of $\overline{Z}$ and the $\overline{CV}_i$. Since $z_0 < 0$ we first compare $\overline{Z}$ to $\overline{CV}_1$. This comparison is

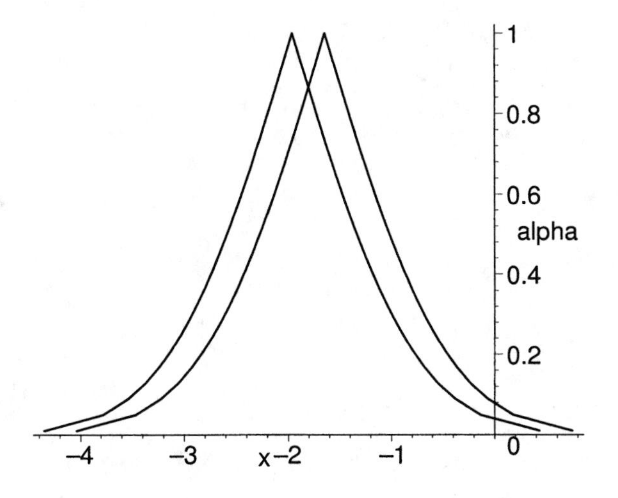

Figure 18.1: Fuzzy Test $\overline{Z}$ verses $\overline{CV}_1$ in Example 18.2.1($\overline{Z}$ right,$\overline{CV}_1$ left)

shown in Figure 18.1. We conclude that $\overline{CV}_1 \approx \overline{Z}$ because the height of their intersection exceeds 0.80. Clearly, $\overline{Z} < \overline{CV}_2$. So we have $\overline{CV}_1 \approx \overline{Z} < \overline{CV}_2$ and there is no decision on H_0. The Maple [2] commands for Figure 18.1 are in Chapter 29. The result of the crisp test would be to not reject H_0.

18.3 References

1. R.V.Hogg and E.A.Tanis: Probability and Statistical Inference, Sixth Edition, Prentice Hall, Upper Saddle River, N.J., 2001.

2. Maple 6, Waterloo Maple Inc., Waterloo, Canada

Chapter 19

Test $d = \mu_1 - \mu_2$, Matched Pairs

19.1 Crisp Test

We begin following Chapter 9. Let $x_1, ..., x_n$ be the values of a random sample from a population Pop I. Let $object_i$, or $person_i$, belong to Pop I which produced measurement x_i in the random sample, $1 \leq i \leq n$. Then, at possibly some later time, we take a second measurement on $object_i$ ($person_i$) and get value y_i, $1 \leq i \leq n$. Then (x_1, y_1),...,(x_n, y_n) are n pairs of dependent measurements. For example, when testing the effectiveness of some treatment for high blood pressure, the x_i are the blood pressure measurements before treatment and the y_i are these measurements on the same person after treatment. The two samples are not independent.

Let $d_i = x_i - y_i$, $1 \leq i \leq n$. Next compute the mean $\overline{d}$ (crisp number, not fuzzy) and the variance s_d^2 of the d_i data. Assume that $n > 30$ so we may use the normal approximation; or assume that the d_i are approximately normally distributed with unknown mean μ_d and unknown variance σ_d^2. Then we wish to perform the following test

$$H_0 : \mu_d = 0, \tag{19.1}$$

verses

$$H_1 : \mu_d \neq 0. \tag{19.2}$$

Our test statistic is

$$t_0 = [\overline{d} - 0]/[s_d/\sqrt{n}], \tag{19.3}$$

which has a t-distribution with $n - 1$ degrees of freedom (Section 8.2 of [1]).

Let γ, $0 < \gamma < 1$, be the significance level of the test. Under the null hypothesis H_0 t_0 has a t-distribution with $n - 1$ degrees of freedom and our decision rule is: (1) reject H_0 if $t_0 \geq t_{\gamma/2}$ or $t_0 \leq -t_{\gamma/2}$; and (2) do not reject

Student	Before(x)	After (y)
1	700	720
2	800	810
3	860	900
4	1070	1080
5	840	840
6	830	810
7	930	950
8	1010	990
9	1100	1100
10	690	700

Table 19.1: SAT Scores in Example 19.2.1

H_0 when $-t_{\gamma/2} < t_0 < t_{\gamma/2}$. The numbers $\pm t_{\gamma/2}$ are called the critical values (cv's) for the test. In the above decision rule $t_{\gamma/2}$ is the t-value so that the probability of a random variable, having a t-distribution with $n-1$ degrees of freedom, exceeding t is $\gamma/2$.

19.2 Fuzzy Model

Now proceed to the fuzzy situation where our statistic for t_0 will become fuzzy $\overline{T}$. In equation (19.3) for t_0 substitute an alpha-cut of our fuzzy estimator $\overline{\mu}_d$ for $\overline{x}$, equation (9.3) of Chapter 9; and substitute an alpha-cut of our fuzzy estimator $\overline{\sigma}_d$ for s_d, the square root of equation (6.10) of Chapter 6. Use interval arithmetic to simplify, assuming all intervals are positive, and we obtain $\overline{T}[\alpha]$ given by equation (13.4) of Chapter 13. Also, $\overline{CV}_2$ ($\overline{CV}_1$) is given in equation (13.12) ((13.13)) in Chapter 13.

Example 19.2.1

Consider the data in Table 19.1 that lists SAT scores on a random sample of ten students before and after the students took a preparatory course for the exam.

We compute $d_i = x_i - y_i$, $1 \le i \le 10$, and then $\overline{d} = -7.00$, $s_d = 18.2878$ and $t_0 = -1.2104$. The hypothesis test is

$$H_0 : \mu_d = 0, \tag{19.4}$$

$$H_1 : \mu_d < 0. \tag{19.5}$$

The claim is that the preparatory course is effective in raising the scores ($\mu_d < 0$). This is a one-sided test. Since this is a one-sided test, see Section

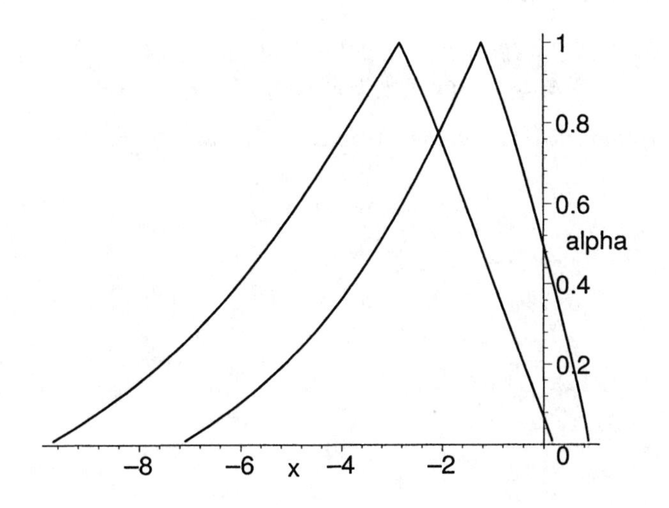

Figure 19.1: Fuzzy Test $\overline{T}$ verses $\overline{CV}_1$ in Example 19.2.1($\overline{T}$ right, $\overline{CV}_1$ left)

12.4 in Chapter 12, we compare $\overline{T}$ with $\overline{CV}_1$. Also, since $\overline{d} < 0$ we will use equation (13.24) for $\overline{T}$, at least for $0 < \alpha^* < \alpha \leq 1$, and equation (13.25) for $\overline{CV}_1$.

Let $\gamma = 0.01$ so that $-t_\gamma = -2.821$ with 9 degrees of freedom. In order to find the Π_i we obtain $\chi^2_{R,0.005} = 23.589$ and $\chi^2_{L,0.005} = 1.735$. Then $L(\lambda) = 23.589 - 14.589\lambda$ and $R(\lambda) = 1.735 + 7.265\lambda$ and

$$\Pi_2 = \sqrt{2.621 - 1.621\lambda}, \tag{19.6}$$

$$\Pi_1 = \sqrt{0.193 + 0.807\lambda}. \tag{19.7}$$

Using $\alpha = f(\lambda)$ from equation (13.14) we may now have Maple [2] do the graphs of $\overline{T}$ and $\overline{CV}_1$.

In comparing $\overline{T}$ and $\overline{CV}_1$ we can see the result in Figure 19.1. The Maple commands for this figure are in Chapter 29. Since $t_0 > -t_{\gamma/2}$ we only need to compare the left side of $\overline{T}$ to the right side of $\overline{CV}_1$. The height of the intersection is $y_0 < 0.8$ (the intersection point is close to 0.8 but it is slightly less than 0.8) and $\overline{T} > \overline{CV}_1$. Hence, we do not reject H_0. The data does not support the claim that the preparatory course raises the scores.

The graphs in Figure 19.1 are correct only to the left of the vertical axis. The right side of $\overline{T}$ goes positive, and we should make the adjustment described in Subsection 13.3.1, but we did not because it will not effect our conclusion.

19.3 References

1. R.V.Hogg and E.A.Tanis: Probability and Statistical Inference, Sixth Edition, Prentice Hall, Upper Saddle River, N.J., 2001.

2. Maple 6, Waterloo Maple Inc., Waterloo, Canada

Chapter 20

Test σ_1^2 verses σ_2^2, Normal Populations

20.1 Crisp Test

We have two populations: Pop I and Pop II. Pop I is normally distributed with unknown mean μ_1 and unknown variance σ_1^2. Pop II is also normally distributed with unknown mean μ_2 and unknown variance σ_2^2. We wish to do the following statistical test

$$H_0 : \sigma_1^2 = \sigma_2^2, \tag{20.1}$$

verses

$$H_1 : \sigma_1^2 \neq \sigma_2^2. \tag{20.2}$$

We collect a random sample of size n_1 from Pop I and let $\overline{x}_1$ be the mean for this data and s_1^2 is the sample variance. We also gather a random sample of size n_2 from Pop II and $\overline{x}_2$ is the mean for the second sample with s_2^2 the variance. We assume these two random samples are independent.

We know that

$$\chi_1^2 = [(n_1 - 1)s_1^2]/\sigma_1^2, \tag{20.3}$$

has a chi-square distribution with $n_1 - 1$ degrees of freedom and

$$\chi_2^2 = [(n_2 - 1)s_2^2]/\sigma_2^2, \tag{20.4}$$

also has a chi-square distribution with $n_1 - 1$ degrees of freedom. The two random variables are independent. Under the null hypothesis (variances equal) our test statistic is (Section 8.3 in [1])

$$f_0 = [\chi_1^2/(n_1 - 1)]/[\chi_2^2/(n_1 - 1)] = s_1^2/s_2^2, \tag{20.5}$$

which has a F distribution with $n_1 - 1$ degrees of freedom for the numerator and $n_2 - 1$ degrees of freedom for the denominator. We will write this as $F(n_1 - 1, n_2 - 1)$.

Let γ, $0 < \gamma < 1$, be the significance level of the test. Our decision rule is : (1) reject H_0 if $f_0 \geq F_{\gamma/2}(n_1 - 1, n_2 - 1)$ or $f_0 \leq F_{1-\gamma/2}(n_1 - 1, n_2 - 1)$; and (2) do not reject H_0 otherwise. In this notation we are using

$$P(F \geq F_s(a, b)) = s. \tag{20.6}$$

where F has an F distribution with a and b degrees of freedom.

20.2 Fuzzy Test

We first fuzzify the test statistic f_0 to $\overline{F}$. Substitute the confidence intervals for the s_i^2, equation (6.10) of Chapter 6, for the s_i^2, $i = 1, 2$, in equation (20.5), and simplify using interval arithmetic (all intervals here are positive) producing

$$\overline{F}[\lambda] = [\Gamma_1(s_1^2/s_2^2), \Gamma_2(s_1^2/s_2^2)], \tag{20.7}$$

for

$$\Gamma_1 = \frac{(n_1 - 1)R_2(\lambda)}{(n_2 - 1)L_1(\lambda)}, \tag{20.8}$$

$$\Gamma_2 = \frac{(n_1 - 1)L_2(\lambda)}{(n_2 - 1)R_1(\lambda)}, \tag{20.9}$$

where: (1) $L_1(\lambda)$ ($R_1(\lambda)$) is given in equation (6.8) ((6.9)) in Chapter 6 using $n_1 - 1$ degrees of freedom; and (2) $L_2(\lambda)$ ($R_2(\lambda)$) is from equation (6.8) ((6.9)) in Chapter 6 having $n_2 - 1$ degrees of freedom. The α-cuts of $\overline{F}$ in this case are functions of λ in $[0, 1]$. We do know that $\lambda = 0$ corresponds to $\alpha = 0.01$ and $\lambda = 1$ is the same as $\alpha = 1$. Otherwise, we get two values of α (assuming $n_1 \neq n_2$) from one value of λ as discussed in Section 17.2 in Chapter 17. We could adopt the compromise solution $\alpha = (\alpha_1 + \alpha_2)/2$ proposed in Section 17.4 of Chapter 17, but instead we will just have λ cuts of $\overline{F}$ and the $\overline{CV}_i$.

The critical region will now be determined by fuzzy critical values $\overline{CV}_i$, $i = 1, 2$. They are determined as in the previous chapters and they are given by their λ-cuts

$$\overline{CV}_1[\lambda] = [\Gamma_1 F_{1-\gamma/2}(n_1 - 1, n_2 - 1), \Gamma_2 F_{1-\gamma/2}(n_1 - 1, n_2 - 1)], \tag{20.10}$$

all α where γ is fixed, and

$$\overline{CV}_2[\lambda] = [\Gamma_1 F_{\gamma/2}(n_1 - 1, n_2 - 1), \Gamma_2 F_{\gamma/2}(n_1 - 1, n_2 - 1)], \tag{20.11}$$

Notice that now, since the F-density is not symmetric with respect to zero, we do not get $\overline{CV}_1 \neq -\overline{CV}_2$.

Now that we have $\overline{F}$, $\overline{CV}_1$ and $\overline{CV}_2$ we may compare $\overline{F}$ and $\overline{CV}_1$ and then compare $\overline{F}$ with $\overline{CV}_2$. The final decision rule was presented in Chapter 12.

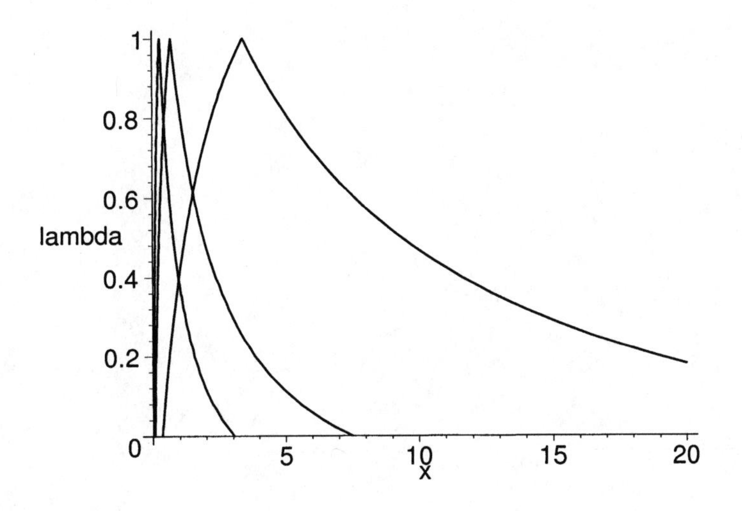

Figure 20.1: Fuzzy Test $\overline{F}$ verses $\overline{CV}_1$ and $\overline{CV}_2$ in Example 20.2.1($\overline{F}$ center, $\overline{CV}_1$ left, $\overline{CV}_2$ right)

Example 20.2.1

The results of the random samples are : (1) $n_1 = 11$, $\overline{x}_1 = 1.03$, $s_1^2 = 0.24$; and (2) $n_2 = 13$, $\overline{x}_2 = 1.66$, $s_2^2 = 0.35$. For $n_1 - 1 = 10$ degrees of freedom we calculate $\chi^2_{L,0.005} = 2.156$, $\chi^2_{R,0.005} = 25.188$ so that

$$L_1(\lambda) = 25.188 - 15.188\lambda, \tag{20.12}$$

$$R_1(\lambda) = 2.156 + 7.844\lambda. \tag{20.13}$$

Next for $n_2 - 1 = 12$ degrees of freedom $\chi^2_{L,0.005} = 3.074$, $\chi^2_{R,0.005} = 28.299$ which implies that

$$L_2(\lambda) = 28.299 - 16.299\lambda, \tag{20.14}$$

$$R_2(\lambda) = 3.074 + 8.926\lambda. \tag{20.15}$$

From these calculations we may find the Γ_i, $i = 1, 2$, and construct the graphs of $\overline{F}$ and the $\overline{CV}_i$. We will use $\gamma = 0.05$ for this test. These figures are shown in Figure 20.1 and their Maple [2] commands are in Chapter 29.

From Figure 20.1 we conclude that $\overline{CV}_1 \approx \overline{F} < \overline{CV}_2$ and there is no decision on H_0. In the crisp case the decision would be do not reject H_0.

Example 20.2.2

Let all the data be the same as in Example 20.2.1 except now $s_1^2 = 0.455$ and $s_2^2 = 0.35$. The graphs are displayed in Figure 20.2. Since the height

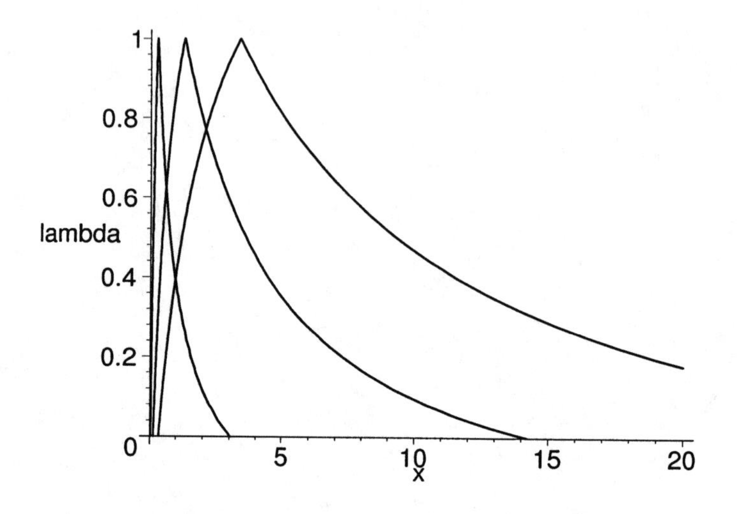

Figure 20.2: Fuzzy Test $\overline{F}$ verses $\overline{CV}_1$ and $\overline{CV}_2$ in Example 20.2.2($\overline{F}$ center, $\overline{CV}_1$ left, $\overline{CV}_2$ right)

on the intersection between $\overline{F}$ and both $\overline{CV}_1$ and $\overline{CV}_2$ is less than 0.80, we conclude $\overline{CV}_1 < \overline{F} < \overline{CV}_2$ and we do not reject H_0. The crisp test produces the same result.

20.3 References

1. R.V.Hogg and E.A.Tanis: Probability and Statistical Inference, Sixth Edition, Prentice Hall, Upper Saddle River, N.J., 2001.

2. Maple 6, Waterloo Maple Inc., Waterloo, Canada

Chapter 21

Fuzzy Correlation

21.1 Introduction

We first present the crisp theory and then the fuzzy results. We have taken the crisp theory from Section 8.8 of [1].

21.2 Crisp Results

Let random variables X and Y have a bivariate normal distribution with parameters μ_x, μ_y, σ_x^2, σ_y^2 and ρ the (linear) correlation coefficient. We want to first estimate ρ and then perform the test

$$H_0 : \rho = 0, \tag{21.1}$$

verses

$$H_1 : \rho \neq 0. \tag{21.2}$$

To estimate ρ we obtain a random sample (x_i, y_i), $1 \leq i \leq n$, from the bivariate distribution. R, the sample correlation coefficient, is the point estimator of ρ and it is computed as follows

$$R = s_{xy}/s_x s_y, \tag{21.3}$$

where

$$s_{xy} = [\sum_{i=1}^{n}(x_i - \overline{x})(y_i - \overline{y})]/(n-1), \tag{21.4}$$

$$s_x = \sqrt{[\sum_{i=1}^{n}(x_i - \overline{x})^2]/(n-1)}, \tag{21.5}$$

$$s_y = \sqrt{[\sum_{i=1}^{n}(y_i - \overline{y})^2]/(n-1)}. \tag{21.6}$$

To obtain confidence interval for ρ we make the transformation

$$W(R) = 0.5 \ln \frac{1+R}{1-R}, \tag{21.7}$$

which has an approximate normal distribution with mean $W(\rho)$ and standard deviation $\sqrt{\frac{1}{n-3}}$. This leads to finding a $(1-\beta)100\%$ confidence interval for ρ. Let the confidence interval be written $[\rho_1(\beta), \rho_2(\beta)]$. Then

$$\rho_1(\beta) = \frac{1+R-(1-R)\exp(s)}{1+R+(1-R)\exp(s)}, \tag{21.8}$$

$$\rho_2(\beta) = \frac{1+R-(1-R)\exp(-s)}{1+R+(1-R)\exp(-s)}, \tag{21.9}$$

$$s = 2z_{\beta/2}/\sqrt{n-3}. \tag{21.10}$$

To perform the hypothesis test we determine the test statistic

$$z_0 = \sqrt{n-3}[W(R) - W(0)], \tag{21.11}$$

because under H_0 $\rho = 0$ and then $W(0) = 0$ also. So, $z_0 = \sqrt{n-3}W(R)$ has an approximate $N(0,1)$ distribution. Let γ, $0 < \gamma < 1$, be the significance level of the test. Usual values for γ are $0.10, 0.05, 0.01$. Now under the null hypothesis H_0 z_0 is $N(0,1)$ and our decision rule is: (1) reject H_0 if $z_0 \geq z_{\gamma/2}$ or $z_0 \leq -z_{\gamma/2}$; and (2) do not reject H_0 when $-z_{\gamma/2} < z_0 < z_{\gamma/2}$.

21.3 Fuzzy Theory

The first thing we want to do is construct our fuzzy estimator for ρ. We just place the $(1-\beta)100\%$ confidence intervals given in equations (21.8)-(21.10), one on top of another to get $\overline{\rho}$.

Example 21.3.1

Let the data be n=16 with computed $R = 0.35$. This is enough to get the graph of the fuzzy estimator $\overline{\rho}$ which is shown in Figure 21.1. The Maple [2] commands for this figure are in Chapter 29.

Next we go to the fuzzy test statistic. Substitute α-cuts of $\overline{\rho}$ for R in equation (21.11), assuming all intervals are positive and simply using interval

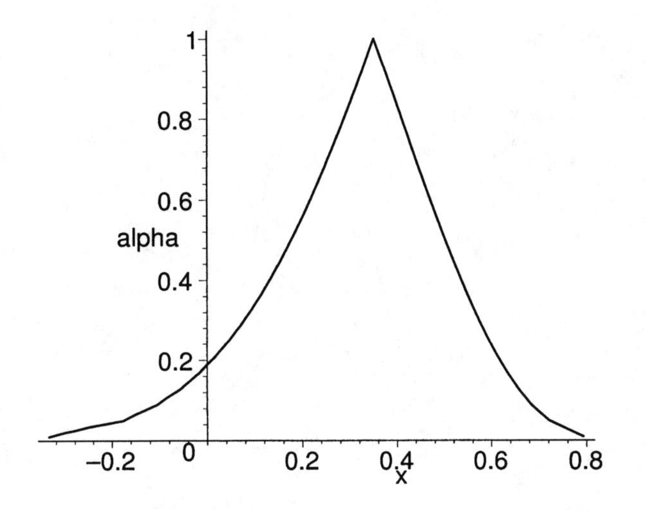

Figure 21.1: Fuzzy Estimator $\overline{\rho}$ in Example 21.3.1

arithmetic, we obtain alpha-cuts of our fuzzy statistic $\overline{Z}$

$$\overline{Z}[\alpha] = \frac{\sqrt{n-3}}{2}[ln\frac{1+\rho_1(\alpha)}{1-\rho_1(\alpha)}, ln\frac{1+\rho_2(\alpha)}{1-\rho_2(\alpha)}]. \tag{21.12}$$

Now substitute the expressions for the $\rho_i(\alpha)$ into the above equation and simplify giving the surprising result

$$\overline{Z}[\alpha] = [z_0 - z_{\alpha/2}, z_0 + z_{\alpha/2}]. \tag{21.13}$$

Using this result we easily determine that

$$\overline{CV}_1[\alpha] = [-z_{\gamma/2} - z_{\alpha/2}, -z_{\gamma/2} + z_{\alpha/2}], \tag{21.14}$$

$$\overline{CV}_2[\alpha] = [z_{\gamma/2} - z_{\alpha/2}, z_{\gamma/2} + z_{\alpha/2}]. \tag{21.15}$$

Example 21.3.2

Use the same data as in Example 21.3.1 and let $\gamma = 0.05$ so that $\pm z_{\gamma/2} = \pm 1.96$ and we compute $z_0 = 1.3176$. The graphs of $\overline{Z}$ and the $\overline{CV}_i$ are shown in Figure 21.2. This figure shows that $\overline{CV}_1 < \overline{Z} < \overline{CV}_2$ since the heights of the intersections are both less than 0.8. Hence we do not reject H_0. The crisp test will produce the same result.

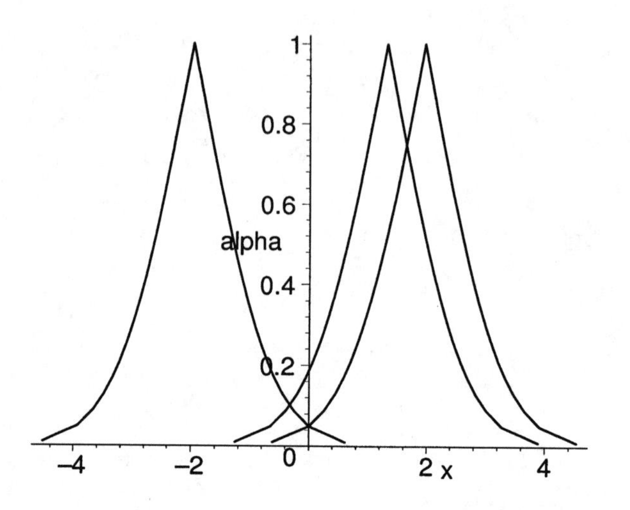

Figure 21.2: Fuzzy Test $\overline{Z}$ verses the $\overline{CV}_i$ in Example 21.3.2($\overline{CV}_1$ left, $\overline{Z}$ middle, $\overline{CV}_2$ right)

If we reject H_0 we believe that we have significant linear correlation between X and Y and then proceed on to the next three chapters to investigate this linear relationship.

21.4 References

1. R.V.Hogg and E.A.Tanis: Probability and Statistical Inference, Sixth Edition, Prentice Hall, Upper Saddle River, N.J., 2001.

2. Maple 6, Waterloo Maple Inc., Waterloo, Canada

Chapter 22

Estimation in Simple Linear Regression

22.1 Introduction

Let us first review the basic theory on crisp simple linear regression. Our development, throughout this chapter and the next two chapters, follows Sections 7.8 and 7.9 in [1]. We have some data (x_i, y_i), $1 \leq i \leq n$, on two variables x and Y. Notice that we start with crisp data and not fuzzy data. Most papers on fuzzy regression assume fuzzy data. The values of x are known in advance and Y is a random variable. We assume that there is no uncertainty in the x data. We can not predict a future value of Y with certainty so we decide to focus on the mean of Y, $E(Y)$. We assume that $E(Y)$ is a linear function of x, say $E(Y) = a + b(x - \overline{x})$. Here $\overline{x}$ is the mean of the x-values and not a fuzzy set. Our model is

$$Y_i = a + b(x_i - \overline{x}) + \epsilon_i, \tag{22.1}$$

where ϵ_i are independent and $N(0, \sigma^2)$ with σ^2 unknown. The basic regression equation for the mean of Y is $y = a + b(x - \overline{x})$ and now we wish to estimate the values of a and b. Notice our basic regression line is not $y = a + bx$, and the expression for the estimator of a will differ between the two models.

We will need the $(1 - \beta)100\%$ confidence interval for a and b. First we require the crisp point estimators of a, b and σ^2. The crisp estimator of a is $\widehat{a} = \overline{y}$ the mean of the y_i values. Next $\widehat{b}$ is $B1/B2$ where

$$B1 = \sum_{i=1}^{n} y_i(x_i - \overline{x}), \tag{22.2}$$

$$B2 = \sum_{i=1}^{n} (x_i - \overline{x})^2. \tag{22.3}$$

Finally

$$\widehat{\sigma}^2 = (1/n)\sum_{i=1}^{n}[y_i - \widehat{a} - \widehat{b}(x_i - \overline{x})]^2. \tag{22.4}$$

Using these expressions we may construct confidence intervals for a and b.

22.2 Fuzzy Estimators

A $(1-\beta)100\%$ confidence interval for a is

$$[\widehat{a} - t_{\beta/2}\sqrt{\widehat{\sigma}^2/(n-2)}, \widehat{a} + t_{\beta/2}\sqrt{\widehat{\sigma}^2/(n-2)}], \tag{22.5}$$

where $t_{\beta/2}$ is the value for a t-distribution, $n-2$ degrees of freedom, so that the probability of exceeding it is $\beta/2$. From this expression we can build the triangular shaped fuzzy number estimator $\overline{a}$ for a by placing these confidence intervals one on top of another.

A $(1-\beta)100\%$ confidence interval for b is

$$[\widehat{b} - t_{\beta/2}\sqrt{C1/C2}, \widehat{b} + t_{\beta/2}\sqrt{C1/C2}], \tag{22.6}$$

where

$$C1 = n\widehat{\sigma}^2, \tag{22.7}$$

and

$$C2 = (n-2)\sum_{i=1}^{n}(x_i - \overline{x})^2. \tag{22.8}$$

These confidence intervals for b will produce the fuzzy number estimator $\overline{b}$ for b.

We will also need the fuzzy estimator for σ^2 in Chapter 24. A $(1-\beta)100\%$ confidence interval for σ^2 is

$$[\frac{n\widehat{\sigma}^2}{\chi^2_{R,\beta/2}}, \frac{n\widehat{\sigma}^2}{\chi^2_{L,\beta/2}}], \tag{22.9}$$

where $\chi^2_{R,\beta/2}$ ($\chi^2_{L,\beta/2}$) is the point on the right (left) side of the χ^2 density where the probability of exceeding (being less than) it is $\beta/2$. This χ^2 distribution has $n-2$ degrees of freedom. Put these confidence intervals together and we obtain $\overline{\sigma}^2$ our fuzzy number estimator of σ^2. However, as discussed in Chapter 6, this fuzzy estimator is biased. It is biased because when we evaluate at $\beta = 1$ we should obtain the point estimator $\widehat{\sigma}^2$ but we do not get this value. So to get an unbiased fuzzy estimator we will define new functions $L(\lambda)$ and $R(\lambda)$ similar to those (equations (6.8) and (6.9)) in Chapter 6. We will employ these definitions of $L(\lambda)$ and $R(\lambda)$ in this chapter and in Chapter 24.

$$L(\lambda) = [1-\lambda]\chi^2_{R,0.005} + \lambda n, \tag{22.10}$$

x	y
70	77
74	94
72	88
68	80
58	71
54	76
82	88
64	80
80	90
61	69

Table 22.1: Crisp Data for Example 22.2.1

$$R(\lambda) = [1-\lambda]\chi^2_{L,0.005} + \lambda n, \tag{22.11}$$

where the degrees of freedom are $n-2$. Then a unbiased $(1-\beta)100\%$ confidence interval for σ^2 is

$$[\frac{n\widehat{\sigma}^2}{L(\lambda)}, \frac{n\widehat{\sigma}^2}{R(\lambda)}], \tag{22.12}$$

for $0 \leq \lambda \leq 1$. If we evaluate this confidence interval at $\lambda = 1$ we obtain $[\widehat{\sigma}^2, \widehat{\sigma}^2] = \widehat{\sigma}^2$. Now β (α) will be a function of λ as shown in equations (6.12) and (6.13) in Chapter 6.

Example 22.2.1

The data set we will use is shown in Table 22.1 which is the data used in Example 7.8-1 in [1]. We will also use this data in the next two chapters. From this data set we compute $\widehat{a} = 81.3$, $\widehat{b} = 0.742$ and $\widehat{\sigma}^2 = 21.7709$.

Then the $(1-\beta)100\%$ confidence interval for a is

$$[81.3 - 1.6496t_{\beta/2}, 81.3 + 1.6496t_{\beta/2}], \tag{22.13}$$

and the same confidence interval for b is

$$[0.742 - 0.1897t_{\beta/2}, 0.742 + 0.1897t_{\beta/2}], \tag{22.14}$$

and the same confidence interval for σ^2 is

$$[\frac{217.709}{L(\lambda)}, \frac{217.709}{R(\lambda)}], \tag{22.15}$$

where

$$L(\lambda) = [1-\lambda](21.955) + 10\lambda, \tag{22.16}$$

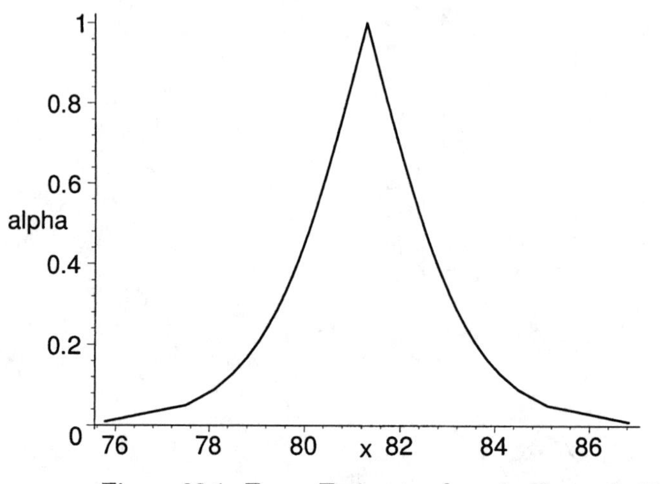

Figure 22.1: Fuzzy Estimator for a in Example 22.2.1

$$R(\lambda) = [1 - \lambda]1.344 + 10\lambda. \tag{22.17}$$

for $0 \leq \lambda \leq 1$. All degrees of freedom are 8. We graphed equations (22.13) and (22.14) as functions of $\beta = \alpha$ for $0.01 \leq \alpha \leq 1$, using Maple [2] and the results are in Figures 22.1 and 22.2. The graph of the fuzzy estimator for the variance, equation (22.15), was done for $\lambda \in [0, 1]$. Maple commands for Figure 22.3 are in Chapter 29. The fuzzy estimator for a (b, σ^2) we shall write as $\overline{a}$ ($\overline{b}$, $\overline{\sigma}^2$).

Once we have these fuzzy estimators of a and b we may go on to fuzzy prediction in the next chapter.

22.3 References

1. R.V.Hogg and E.A.Tanis: Probability and Statistical Inference, Sixth Edition, Prentice Hall, Upper Saddle River, N.J., 2001.

2. Maple 6, Waterloo Maple Inc., Waterloo, Canada.

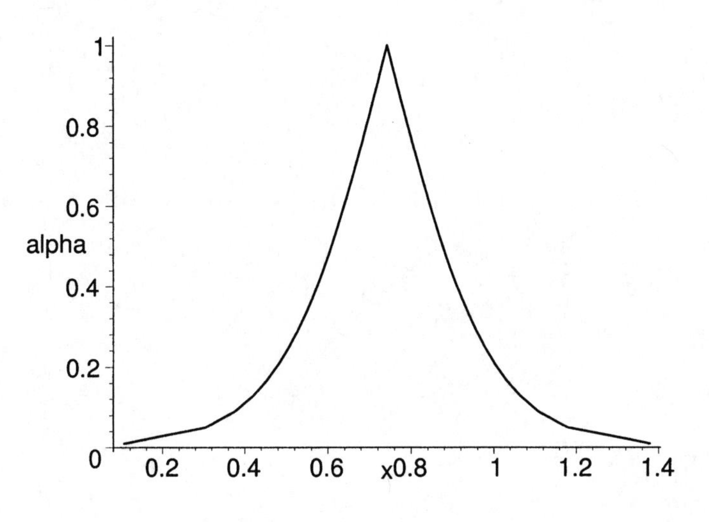

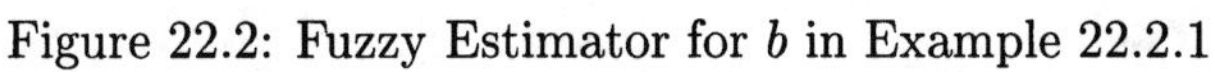
Figure 22.2: Fuzzy Estimator for b in Example 22.2.1

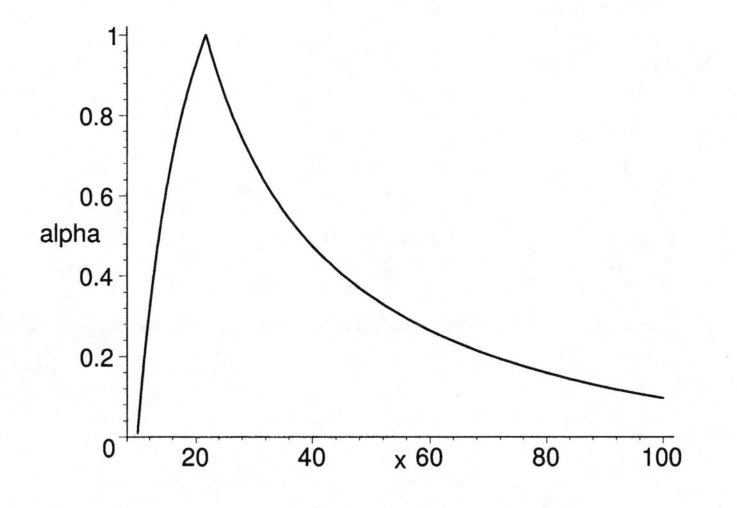

Figure 22.3: Fuzzy Estimator for σ^2 in Example 22.2.1

Chapter 23

Fuzzy Prediction in Linear Regression

23.1 Prediction

From the previous chapter we have our fuzzy regression equation

$$\overline{y}(x) = \overline{a} + \overline{b}(x - \overline{x}), \tag{23.1}$$

for $\overline{y}(x)$, with $\overline{a}$ and $\overline{b}$ fuzzy numbers and x and $\overline{x}$ real numbers. $\overline{y}(x)$ is our fuzzy number estimator for the mean of Y ($E(Y)$) given x, and we show this dependence on x with the notation $\overline{y}(x)$. Now $\overline{x}$ is a fixed real number but we may choose new values for x to predict new fuzzy values for $E(Y)$.

Let $\overline{a}[\alpha] = [a_1(\alpha), a_2(\alpha)]$, $\overline{b}[\alpha] = [b_1(\alpha), b_2(\alpha)]$ and $\overline{y}(x)[\alpha] = [y(x)_1(\alpha), y(x)_2(\alpha)]$. All fuzzy calculations will be done using α-cuts and interval arithmetic. The main thing to remember now from interval arithmetic (Chapter 2, Section 2.3.2) is that $c[a, b]$ equals $[ca, cb]$ if $c > 0$ but it is $[cb, ca]$ when $c < 0$. Then

$$y(x)_1(\alpha) = a_1(\alpha) + (x - \overline{x})b_1(\alpha), \tag{23.2}$$

when $(x - \overline{x}) > 0$ and

$$y(x)_1(\alpha) = a_1(\alpha) + (x - \overline{x})b_2(\alpha), \tag{23.3}$$

if $(x - \overline{x}) < 0$. Similarly

$$y(x)_2(\alpha) = a_2(\alpha) + (x - \overline{x})b_2(\alpha), \tag{23.4}$$

when $(x - \overline{x}) > 0$ and

$$y(x)_2(\alpha) = a_2(\alpha) + (x - \overline{x})b_1(\alpha), \tag{23.5}$$

if $(x - \overline{x}) < 0$. The alpha-cuts of $\overline{a}$ and $\overline{b}$ were determined in the previous chapter. There the α-cut is the $(1 - \alpha)100\%$ confidence interval.

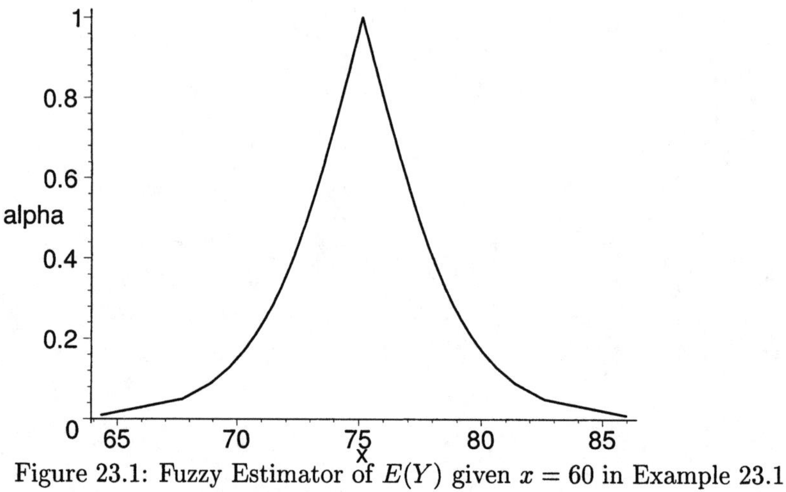

Figure 23.1: Fuzzy Estimator of $E(Y)$ given $x = 60$ in Example 23.1

Example 23.1

We use the same data as in Example 22.2.1 in Chapter 22. Here we will find $\overline{y}(60)$ and $\overline{y}(70)$. Notice that we are using $x = 70$ which is already in the data set in Table 22.1. First consider $x = 60$. Then $(x - \overline{x}) = -8.3 < 0$ because $\overline{x} = 68.3$. We use equations (23.3) and (23.5). Using Maple [2] the graph of $\overline{y}(60)$ is in Figure 23.1. If $x = 70$ then $(x - \overline{x}) = 1.7 > 0$ and use equations (23.2) and (23.4). The graph of $\overline{y}(70)$ is shown in Figure 23.2. The Maple commands for the $x = 60$ case are in Chapter 29.

Now let us compare these results to those obtained from probability theory. First $\overline{y}(x)[0]$ is like a 99% confidence interval for $y(x)$ because it uses $\overline{a}[0]$ ($\overline{b}[0]$) which is a 99% confidence interval for a (b). So we will compare these $\alpha = 0$ cuts to: (1) the 99% confidence interval for the mean of Y ($E(Y)$) given $x = 60(70)$; and (2) the 99% confidence interval for a value of y given $x = 60$ (70). Expressions for both of these crisp confidence intervals may be found in the statistics book [1], so we will not reproduce them here. The results are in Table 23.1 where "CI" denotes "confidence interval". Notice that in Table 23.1: (1) the 99% confidence interval for $E(Y)$ is a subset of $y(x)[0]$ for both $x = 60$ and $x = 70$; and (2) the 99% confidence interval for a value of y contains the interval $y(x)[0]$ for $x = 60, 70$. We know from crisp statistics that the confidence interval for $E(Y)$ will always be a subset of the confidence interval for a value of y. However, we do not always expect, for all other data sets, $y(x)[0]$ to be between these other two intervals.

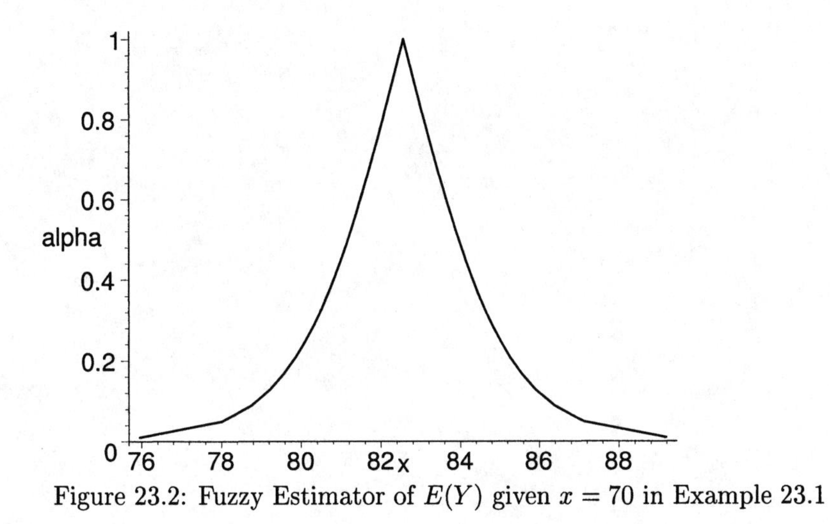

Figure 23.2: Fuzzy Estimator of $E(Y)$ given $x = 70$ in Example 23.1

Confidence Interval	$x = 60$	$y = 70$
$\overline{y}(x)[0]$	$[64.32, 85.96]$	$[75.94, 89.18]$
99% CI for $E(Y)$	$[67.49, 82.79]$	$[76.92, 88.20]$
99% CI for y	$[56.04, 94.24]$	$[64.17, 100.95]$

Table 23.1: Comparing the 99% Confidence Intervals in Example 23.1

23.2 References

1. R.V.Hogg and E.A.Tanis: Probability and Statistical Inference, Sixth Edition, Prentice Hall, Upper Saddle River, N.J., 2001.

2. Maple 6, Waterloo Maple Inc., Waterloo, Canada.

Chapter 24

Hypothesis Testing in Regression

24.1 Introduction

We look at two fuzzy hypothesis tests in this chapter: (1) in the next section $H_0 : a = a_0$ verses $H_1 : a > a_0$ a one-sided test; and (2) in the following section $H_0 : b = 0$ verses $H_1 : b \neq 0$. In both cases we first review the crisp (non-fuzzy) test before the fuzzy test. The non-fuzzy hypothesis tests are based on Sections 7.8 and 7.9 of [1].

24.2 Tests on a

Let us first review the crisp situation. We wish to do the following hypothesis test

$$H_0 : a = a_0, \tag{24.1}$$

verses

$$H_1 : a > a_0, \tag{24.2}$$

which is a one-sided test. Then we determine the statistic

$$t_0 = \frac{\widehat{a} - a_0}{\sqrt{\widehat{\sigma}^2/(n-2)}}, \tag{24.3}$$

which, under H_0, has a t distribution with $(n-2)$ degrees of freedom. Let γ, $0 < \gamma < 1$, be the significance level of the test. Usual values for γ are $0.10, 0.05, 0.01$. Our decision rule is: (1) reject H_0 if $t_0 \geq t_\gamma$; and (2) do not reject H_0 when $t_0 < t_\gamma$. In the above decision rule t_γ is the t-value so that the probability of a random variable, having the t probability density, exceeding t is γ. The critical region is $[t_\gamma, \infty)$ with critical value t_γ.

Now proceed to the fuzzy situation where our fuzzy estimator of a is the triangular shaped fuzzy number $\overline{a}$ developed in Chapter 22. We will also need the fuzzy estimator for σ^2 also given in Chapter 22. Then our fuzzy statistic is

$$\overline{T} = \frac{\overline{a} - a_0}{\sqrt{\overline{\sigma}^2/(n-2)}}. \tag{24.4}$$

All fuzzy calculations will be performed via α-cuts and interval arithmetic. We find, after substituting the intervals for an alpha-cut of $\overline{a}$ and $\overline{\sigma}^2$ into the expression for $\overline{T}$, using interval arithmetic, and simplification, that

$$\overline{T}[\alpha] = [\Pi_1(t_0 - t_{\alpha/2}), \Pi_2(t_0 + t_{\alpha/2})], \tag{24.5}$$

where

$$\Pi_1 = \sqrt{R(\lambda)/n}, \tag{24.6}$$

and

$$\Pi_2 = \sqrt{L(\lambda)/n}. \tag{24.7}$$

$L(\lambda)$ and $R(\lambda)$ were defined in equations (22.10) and (22.11), respectively, in Chapter 22.

We have assumed that all intervals are positive in the derivation of equation (24.5). The interval for an alpha-cut of $\overline{a}$ may be positive or negative, but the interval for an alpha-cut of $\overline{\sigma}^2$ is always positive. When the left end point (or both end points) of the interval for an alpha-cut of $\overline{a}$ is negative we have to make some changes in equation (24.5). See section 13.3.1 in Chapter 13 for the details.

Now that we know the alpha-cuts of the fuzzy statistic we can find α-cuts of the fuzzy critical value $\overline{CV}_2$ for this one-sided test. As in previous chapters we get

$$\overline{CV}_2[\alpha] = [\Pi_1(t_\gamma - t_{\alpha/2}), \Pi_2(t_\gamma + t_{\alpha/2})]. \tag{24.8}$$

In this equation γ is fixed and alpha varies in the interval $[0.01, 1]$.

We now have a fuzzy set $\overline{T}$ for our test statistic and a fuzzy set $\overline{CV}_2$ for the critical value. Our final decision will depend on the relationship between $\overline{T}$ and $\overline{CV}_2$. Our test becomes : (1) reject H_0 if $\overline{T} > \overline{CV}_2$; (2) do not reject if $\overline{T} < \overline{CV}_2$; and (3) there is no decision on H_0 if $\overline{T} \approx \overline{CV}_2$.

Example 24.2.1

We will still use the data in Table 22.1 and we have computed $\widehat{a} = 81.3$, $\widehat{b} = 0.742$ and $\widehat{\sigma}^2 = 21.7709$ with $n = 10$. Let $\gamma = 0.05$, $a_0 = 80$ and determine $t_0 = 0.7880$ and $t_{0.05} = 1.860$ with 8 degrees of freedom. We compute

$$L(\lambda) = 21.955 - 11.955\lambda, \tag{24.9}$$

$$R(\lambda) = 1.344 + 8.656\lambda, \tag{24.10}$$

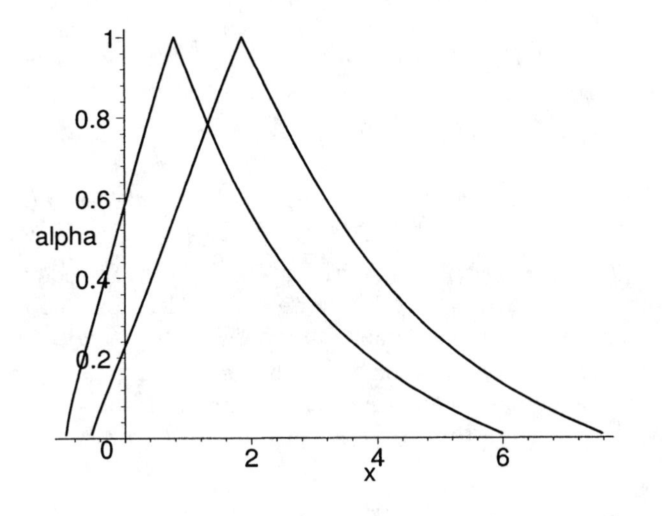

Figure 24.1: Fuzzy Test $\overline{T}$ verses $\overline{CV}_2$ in Example 24.2.1($\overline{T}$ left, $\overline{CV}_2$ right)

$$\Pi_1 = \sqrt{0.1344 + 0.8656\lambda}, \tag{24.11}$$

$$\Pi_2 = \sqrt{2.1955 - 1.1955\lambda}. \tag{24.12}$$

From these results we may get the graphs of $\overline{T}$ and $\overline{CV}_2$ and they are shown in Figure 24.1. The Maple [2] commands for this figure are in Chapter 29.

From Figure 24.1 we see that $\overline{T} < \overline{CV}_2$ since the height of the intersection is less than 0.8. We therefore conclude: do not reject H_0. Of course, the crisp test would have the same result.

The graph of $\overline{T}$ in Figure 24.1 is not entirely correct on its left side to the left of the vertical axis. This is because in computing α-cuts of $\overline{T}$ the left end point of the interval for the numerator (see Section 13.3.1) goes negative for $0 \leq \alpha < \alpha^*$ and we had assumed it was always positive. Also, this effects the left side of $\overline{CV}_2$ for $0 \leq \alpha < \alpha^*$. However, these changes do not effect the final decision because it depends on comparing the right side of $\overline{T}$ to the left side of $\overline{CV}_2$.

24.3 Tests on b

Let us first discuss the crisp hypothesis test. We wish to do the following hypothesis test

$$H_0 : b = 0, \tag{24.13}$$

verses

$$H_1 : b \neq 0, \tag{24.14}$$

which is a two-sided test. Next determine the statistic

$$t_0 = \frac{\widehat{b} - 0}{\sqrt{d\widehat{\sigma}^2/(n-2)}}, \tag{24.15}$$

where

$$d = \frac{n}{\sum_{i=1}^{n}(x_i - \overline{x})^2}, \tag{24.16}$$

which, under H_0, t_0 has a t distribution with $(n-2)$ degrees of freedom. Let γ, $0 < \gamma < 1$, be the significance level of the test. Our decision rule is: (1) reject H_0 if $t_0 \geq t_{\gamma/2}$ or if $t_0 \leq -t_{\gamma/2}$; and (2) otherwise do not reject H_0.

Now proceed to the fuzzy situation where our estimate of b is the triangular shaped fuzzy number $\overline{b}$ and our fuzzy estimator $\overline{\sigma}^2$ of σ^2 is also a fuzzy number. These fuzzy estimators were deduced in Chapter 22. Then our fuzzy statistic is

$$\overline{T} = \frac{\overline{b} - 0}{\sqrt{d\overline{\sigma}^2/(n-2)}}. \tag{24.17}$$

All fuzzy calculations will be performed via α-cuts and interval arithmetic. We find, after substituting the intervals for an alpha-cuts of $\overline{b}$ and $\overline{\sigma}^2$ into the expression for $\overline{T}$, using interval arithmetic, assuming all intervals are positive, that

$$\overline{T}[\alpha] = [\Pi_1(t_0 - t_{\alpha/2}), \Pi_2(t_0 + t_{\alpha/2})], \tag{24.18}$$

where

$$\Pi_1 = \sqrt{R(\lambda)/n}, \tag{24.19}$$

and

$$\Pi_2 = \sqrt{L(\lambda)/n}. \tag{24.20}$$

$L(\lambda)$ and $R(\lambda)$ were defined in equations (22.10) and (22.11) in Chapter 22.

Now that we know the alpha-cuts of the fuzzy statistic we can find α-cuts of the fuzzy critical values $\overline{CV}_i$, $i = 1, 2$. As in previous chapters we obtain

$$\overline{CV}_1[\alpha] = [\Pi_2(-t_{\gamma/2} - t_{\alpha/2}), \Pi_1(-t_{\gamma/2} + t_{\alpha/2})], \tag{24.21}$$

and because $\overline{CV}_2 = -\overline{CV}_1$

$$\overline{CV}_2[\alpha] = [\Pi_1(t_{\gamma/2} - t_{\alpha/2}), \Pi_2(t_{\gamma/2} + t_{\alpha/2})]. \tag{24.22}$$

In these equations γ is fixed and alpha varies in the interval [0.01, 1].

Given the fuzzy numbers $\overline{T}$ and the $\overline{CV}_i$, $i = 1, 2$, we may compare $\overline{T}$ to $\overline{CV}_1$ and then to $\overline{CV}_2$ to determine our final conclusion on H_0.

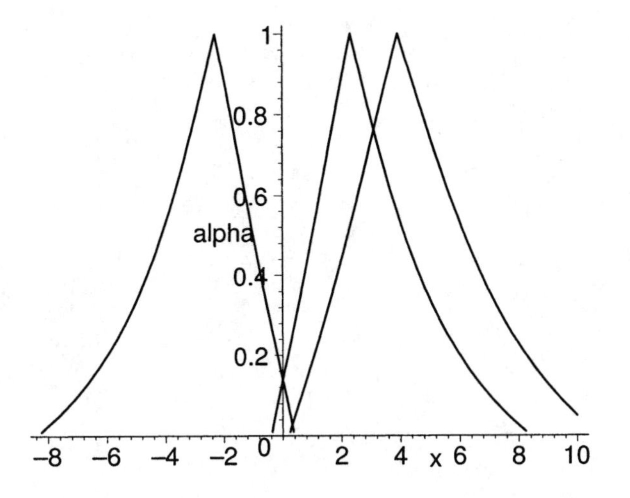

Figure 24.2: Fuzzy Test $\overline{T}$ verses the $\overline{CV}_i$ in Example 24.3.1 ($\overline{CV}_1$ left, $\overline{CV}_2$ center, $\overline{T}$ right)

Example 24.3.1

We will still use the data in Table 22.1 and we have already computed $\widehat{b} = 0.742$, and $\widehat{\sigma}^2 = 21.7709$ with $n = 10$. Let $\gamma = 0.05$, and compute $t_0 = 3.9111$ and $t_{0.025} = 2.306$ with 8 degrees of freedom.

The values of $L(\lambda)$, $R(\lambda)$, Π_1 and Π_2 are all the same as in Example 24.2.1. All that has changed is the value of t_0 and that now we use both $\overline{CV}_1$ and $\overline{CV}_2$ for a two-sided test.

The graphs of $\overline{T}$ and the $\overline{CV}_i$ are shown in Figure 24.2. It is evident that $\overline{CV}_2 < \overline{T}$, because the height of the intersection is less than 0.8. Hence we reject H_0, the same as in the crisp case.

24.4 References

1. R.V.Hogg and E.A.Tanis: Probability and Statistical Inference, Sixth Edition, Prentice Hall, Upper Saddle River, N.J., 2001.

2. Maple 6, Waterloo Maple Inc., Waterloo, Canada.

Chapter 25

Estimation in Multiple Regression

25.1 Introduction

We first review the basic theory on crisp multiple linear regression. For simplicity let us work with only two independent variables x_1 and x_2. Our development, throughout this chapter, and the next two chapters, follows [1]. We have some data (x_{1i}, x_{2i}, y_i), $1 \leq i \leq n$, on three variables x_1, x_2 and Y. Notice that we start with crisp data and not fuzzy data. The values of x_1 and x_2 are known in advance and Y is a random variable. We assume that there is no uncertainty in the x_1 and x_2 data. We can not predict a future value of Y with certainty so we decide to focus on the mean of Y, $E(Y)$. We assume that $E(Y)$ is a linear function of x_1 and x_2, say $E(Y) = a + bx_1 + cx_2$. Our model is

$$Y_i = a + bx_{1i} + cx_{2i} + \epsilon_i, \tag{25.1}$$

$1 \leq i \leq n$. The basic regression equation for the mean of Y is $y = a + bx_1 + cx_2$ and now we wish to estimate the values of a, b and c.

We will need the $(1-\beta)100\%$ confidence interval for a, b and c. First, we require the crisp point estimators of a, b, c and σ^2. It is best now to turn to matrix notation in order to describe the estimators and their confidence intervals.

Let vector $\theta = [a, b, c]$, vector $\epsilon = [\epsilon_1, ..., \epsilon_n]$ and vector $y = [y_1, ..., y_n]$. If v is a $1 \times n$ vector we will use the notation v^t for the transpose of v. Then v^t is a $n \times 1$ vector. Define the $n \times 3$ matrix X as

$$X = \begin{pmatrix} 1 & x_{11} & x_{21} \\ 1 & x_{12} & x_{22} \\ .. & .. & .. \\ .. & .. & .. \\ 1 & x_{1n} & x_{2n} \end{pmatrix}. \tag{25.2}$$

Next define $\widehat{\theta} = [\widehat{a}, \widehat{b}, \widehat{c}]$ the vector of point estimates of a, b, c. Then

$$\widehat{\theta}^t = (X^t X)^{-1} X y^t, \tag{25.3}$$

which gives $\widehat{a}$,$\widehat{b}$ and $\widehat{c}$.

The distribution of ϵ is also needed. We know that the expected value of $\epsilon^t \epsilon$ is $\sigma^2 I$ for 3×3 identity matrix I and unknown variance σ^2. In the next two chapters a point estimate for σ^2, and confidence intervals, are required. A point estimator $\widehat{\sigma}^2$ for σ^2 is

$$\widehat{\sigma}^2 = \sum_{i=1}^{n} e_i^2/(n-3), \tag{25.4}$$

where

$$\sum_{i=1}^{n} e_i^2 = \sum_{i=1}^{n} (y_i - \widehat{y}_i)^2, \tag{25.5}$$

$$\widehat{y}_i = \widehat{a} + \widehat{b} x_{1i} + \widehat{c} x_{2i}. \tag{25.6}$$

25.2 Fuzzy Estimators

Now we may find the confidence intervals for a, b, c and σ^2. Let $(X^t X)^{-1} = A = [a_{ij}]$. A $(1-\beta)100\%$ confidence interval for a is

$$[\widehat{a} - t_{\beta/2}\widehat{\sigma}\sqrt{a_{11}}, \widehat{a} + t_{\beta/2}\widehat{\sigma}\sqrt{a_{11}}], \tag{25.7}$$

for a_{11} the first element along the main diagonal of matrix A. Then a $(1-\beta)100\%$ confidence interval for b is

$$[\widehat{b} - t_{\beta/2}\widehat{\sigma}\sqrt{a_{22}}, \widehat{a} + t_{\beta/2}\widehat{\sigma}\sqrt{a_{22}}], \tag{25.8}$$

and for c

$$[\widehat{c} - t_{\beta/2}\widehat{\sigma}\sqrt{a_{33}}, \widehat{a} + t_{\beta/2}\widehat{\sigma}\sqrt{a_{33}}]. \tag{25.9}$$

In the t distribution, to find the critical value $t_{\beta/2}$, we use $n-3$ degrees of freedom. Now put these confidence intervals together, one on top of another, to get the fuzzy estimators $\overline{a}$, $\overline{b}$, $\overline{c}$ of a, b, c, respectively.

The next item we need is a confidence interval for σ^2. It is known that

$$(n-3)\widehat{\sigma}^2/\sigma^2, \tag{25.10}$$

has a chi-square distribution with $n-3$ degrees of freedom. Then

$$P(\chi^2_{L,\beta/2} < \frac{(n-3)\widehat{\sigma}^2}{\sigma^2} < \chi^2_{R,\beta/2}) = 1 - \beta, \tag{25.11}$$

and if we solve this equation for σ^2, it leads to the $(1-\beta)100\%$ confidence interval for σ^2, which is

$$[\frac{(n-3)\widehat{\sigma}^2}{\chi^2_{R,\beta/2}}, \frac{(n-3)\widehat{\sigma}^2}{\chi^2_{L,\beta/2}}]. \tag{25.12}$$

where $\chi^2_{R,\beta/2}$ ($\chi^2_{L,\beta/2}$) is the point on the right (left) side of the χ^2 density where the probability of exceeding (being less than) it is $\beta/2$. This χ^2 distribution has $n-3$ degrees of freedom. Put these confidence intervals together and we obtain $\overline{\sigma}^2$ our fuzzy number estimator of σ^2. However, as discussed in Chapter 6 and 22, this fuzzy estimator is biased. It is biased because when we evaluate at $\beta = 1$ we should obtain the point estimator $\widehat{\sigma}^2$ but we do not get this value. So to get an unbiased fuzzy estimator we will define new functions $L(\lambda)$ and $R(\lambda)$, similar to those in Chapter 6. We will employ these definitions of $L(\lambda)$ and $R(\lambda)$ in this chapter and in Chapter 27.

$$L(\lambda) = [1-\lambda]\chi^2_{R,0.005} + \lambda(n-3), \tag{25.13}$$

$$R(\lambda) = [1-\lambda]\chi^2_{L,0.005} + \lambda(n-3), \tag{25.14}$$

where the degrees of freedom are $n-3$. Then a unbiased $(1-\beta)100\%$ fuzzy estimator for σ^2 is $\overline{\sigma}^2$ whose α-cuts are

$$[\frac{(n-3)\widehat{\sigma}^2}{L(\lambda)}, \frac{(n-3)\widehat{\sigma}^2}{R(\lambda)}], \tag{25.15}$$

for $0 \leq \lambda \leq 1$. If we evaluate this confidence interval at $\lambda = 1$ we obtain $[\widehat{\sigma}^2, \widehat{\sigma}^2] = \widehat{\sigma}^2$. Now β will be a function of λ as shown in equations (6.12) and (6.13) in Chapter 6.

Example 25.2.1

The data we shall use is in Table 25.1 which is from an example in [1]. This same data will be in the examples in the next two chapters. We want to construct the graphs of the fuzzy estimators.

We first, using Maple [2], computed the point estimators and obtained $\widehat{a} = -49.3413$, $\widehat{b} = 1.3642$, $\widehat{c} = 0.1139$ and $\widehat{\sigma}^2 = 12.9236$. Next we found the values down the main diagonal of $(X^tX)^{-1}$ and they were $a_{11} = 44.7961$, $a_{22} = 0.001586$ and $a_{33} = 0.001591$. The equations that determine the alpha-cuts of the fuzzy estimators are

$$-49.3413 \pm t_{\alpha/2}(24.0609), \tag{25.16}$$

for $\overline{a}$, and for $\overline{b}$ it is

$$1.3642 \pm t_{\alpha/2}(0.1432), \tag{25.17}$$

Y	x_1	x_2
100	100	100
106	104	99
107	106	110
120	111	126
110	111	113
116	115	103
123	120	102
133	124	103
137	126	98

Table 25.1: Crisp Data for Example 25.2.1

and

$$0.1139 \pm t_{\alpha/2}(0.1434), \tag{25.18}$$

goes with $\overline{c}$, and

$$[\frac{77.5416}{L(\lambda)}, \frac{77.5416}{R(\lambda)}], \tag{25.19}$$

for $\overline{\sigma}^2$. These are all graphed, $0.01 \leq \alpha \leq 1$, and the results are in Figures 25.1-25.4.

25.3 References

1. J.Johnston: Econometric Methods, Second Edition, McGraw-Hill, N.Y., 1972.

2. Maple 6, Waterloo Maple Inc., Waterloo, Canada.

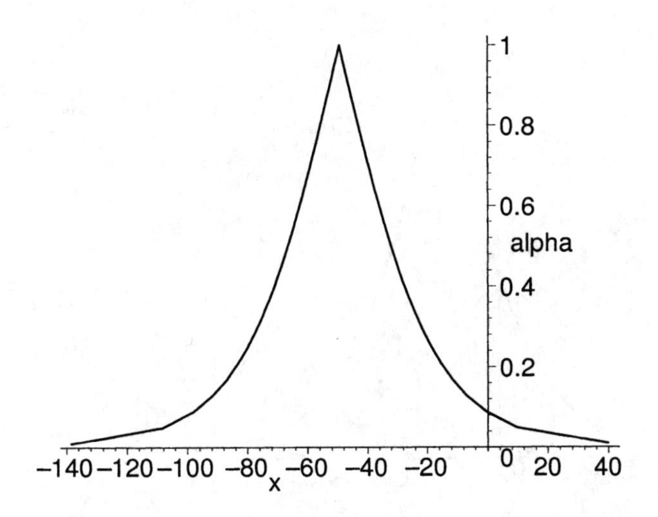

Figure 25.1: Fuzzy Estimator $\overline{a}$ for a in Example 25.2.1

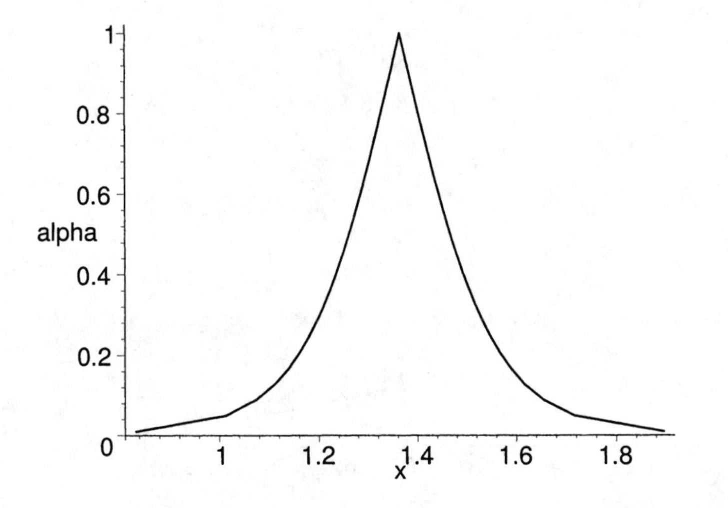

Figure 25.2: Fuzzy Estimator $\overline{b}$ for b in Example 25.2.1

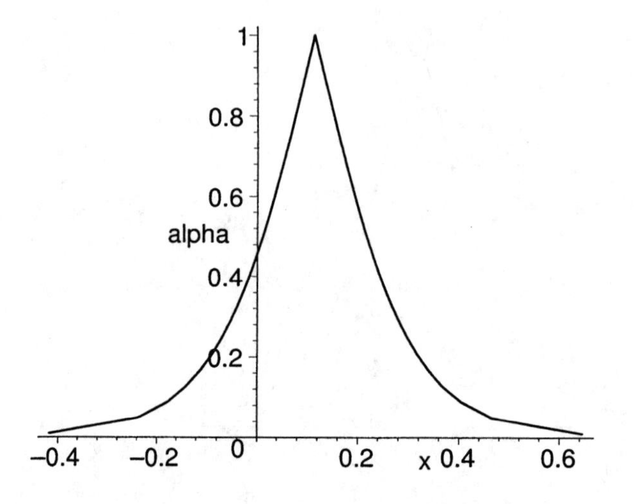

Figure 25.3: Fuzzy Estimator $\overline{c}$ for c in Example 25.2.1

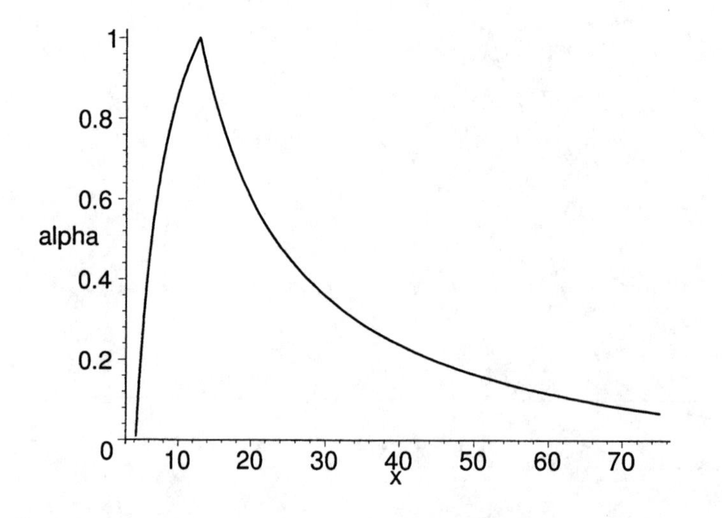

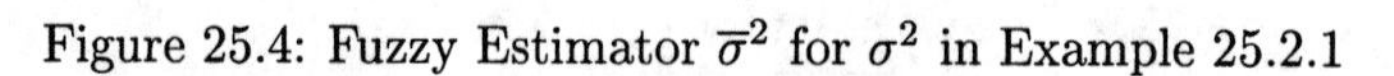

Figure 25.4: Fuzzy Estimator $\overline{\sigma}^2$ for σ^2 in Example 25.2.1

Chapter 26

Fuzzy Prediction in Regression

26.1 Prediction

From the previous chapter we have our fuzzy regression equation

$$\overline{y}(x_1, x_2) = \overline{a} + \overline{b}x_1 + \overline{c}x_2, \tag{26.1}$$

for $\overline{y}(x_1, x_2)$, with $\overline{a}$, $\overline{b}$ and $\overline{c}$ fuzzy numbers and x_1, x_2 real numbers. $\overline{y}(x_1, x_2)$ is our fuzzy number estimator for the mean of Y ($E(Y)$) given x_1 and x_2, and we show this dependence on x_1 and x_2 with the notation $\overline{y}(x_1, x_2)$. We may choose new values for x_1 and x_2 to predict new fuzzy values for $E(Y)$.

Let $\overline{a}[\alpha] = [a_1(\alpha), a_2(\alpha)]$, $\overline{b}[\alpha] = [b_1(\alpha), b_2(\alpha)]$, $\overline{c}[\alpha] = [c_1(\alpha), c_2(\alpha)]$ and $\overline{y}(x_1, x_2)[\alpha] = [y(x_1, x_2)_1(\alpha), y(x_1, x_2)_2(\alpha)]$. All fuzzy calculations will be done using α-cuts and interval arithmetic. Now from Example 25.2.1, and the data in Table 25.1, we assume the new values of x_1 and x_2 are positive. The only thing to remember from interval arithmetic (Chapter 2, Section 2.3.2) is that $e[a, b]$ equals $[ea, eb]$ when $e > 0$. Then

$$y(x_1, x_2)_1(\alpha) = a_1(\alpha) + x_1 b_1(\alpha) + x_2 c_1(\alpha), \tag{26.2}$$

and

$$y(x_1, x_2)_2(\alpha) = a_2(\alpha) + x_1 b_2(\alpha) + x_2 c_2(\alpha), \tag{26.3}$$

for all $\alpha \in [0, 1]$. The alpha-cuts of $\overline{a}$, $\overline{b}$ and $\overline{c}$ were determined in the previous chapter. There the α-cut is the $(1 - \alpha)100\%$ confidence interval.

Example 26.1.1

We use the same data as in Example 25.2.1 in Chapter 25. Let us assume now that the data in Table 25.1 is yearly data with the last row corresponding to 2003. Assuming values for x_1 and x_2 for 2004 and 2005 we wish to predict $E(Y)$ for those two future years. We will find $\overline{y}(128, 96)$ and $\overline{y}(132, 92)$.

First we graphed equations (26.2) and (26.3) using $x_1 = 128$ and $x_2 = 96$. The result is in Figure 26.1. The Maple [2] commands for this figure are in Chapter 29. Next we graphed these two equations having $x_1 = 132$ and $x_2 = 92$ which is shown in Figure 26.2.

Now let us compare these results to those obtained from probability theory. First $\overline{y}(x_1, x_2)[0]$ is like a 99% confidence interval for $y(x_1, x_2)$ because it uses $\overline{a}[0]$ ($\overline{b}[0]$, $\overline{c}[0]$) which is a 99% confidence interval for a (b, c). So we will compare these $\alpha = 0$ cuts to: (1) the 99% confidence interval for the mean of Y ($E(Y)$) given $x_1 = 128$, $x_2 = 96$($x_1 = 132$,$x_2 = 92$); and (2) the 99% confidence interval for a value of y given $x_1 = 128$, $x_2 = 96$ ($x_1 = 132$, $x_2 = 92$). Expressions for both of these crisp confidence intervals may be found in Section 5-5 of [1] and are reproduced below.

Let $c = (1, x_1^*, x_2^*)$ where x_i^* are new values of x_i, $i = 1, 2$. The 99% confidence interval for $E(Y)$ is

$$c\widehat{\theta}^t \pm (3.707)\widehat{\sigma}\sqrt{c(X^tX)^{-1}c^t}, \tag{26.4}$$

where $t_{0.005} = 3.707$ for 6 degrees of freedom and the rest of the terms ($\widehat{\theta}$, $\widehat{\sigma}$,..) were defined in Chapter 25. The 99% confidence interval for the value of y is

$$c\widehat{\theta}^t \pm (3.707)\widehat{\sigma}\sqrt{1 + c(X^tX)^{-1}c^t}. \tag{26.5}$$

The results are in Table 26.1 where "CI" denotes "confidence interval". Notice that in Table 26.1 that: (1) the 99% confidence interval for $E(Y)$ is a subset of $\overline{y}(x_1, x_2)[0]$ for both $x_1 = 128$, $x_2 = 96$ and $x_1 = 132$, $x_2 = 92$; and (2) the 99% confidence interval for a value of y is also contained in the interval $\overline{y}(x_1, x_2)[0]$ for the given new values of x_1 and x_2. We know from crisp statistics that the confidence interval for $E(Y)$ will always be a subset of the confidence interval for a value of y. However, we do not always expect, for all other data sets, $\overline{y}(x_1, x_2)[0]$ will contain the other two intervals (see Example 23.1). If fact, in this example the interval $\overline{y}(x_1, x_2)[0]$ turns out to be rather large because it combines three intervals $\overline{a}[0]$, $128(132)\overline{b}[0]$ and $96(92)\overline{c}[0]$. For example, using $x_1 = 128$ and $x_2 = 96$, then: (1) $\overline{a}[0]$ has length ≈ 178; (2) $128\overline{b}[0]$ has length ≈ 136; and (3) $96\overline{c}[0]$ has length ≈ 102. We add these lengths up we get that the length of the interval for $\overline{y}(128, 96)[0]$ is ≈ 416.

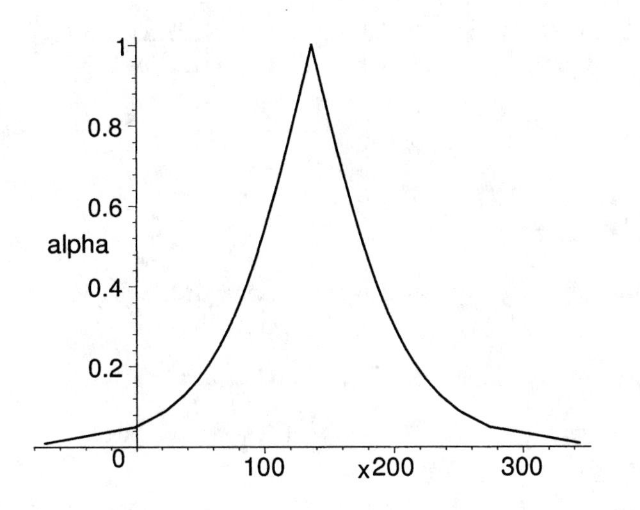

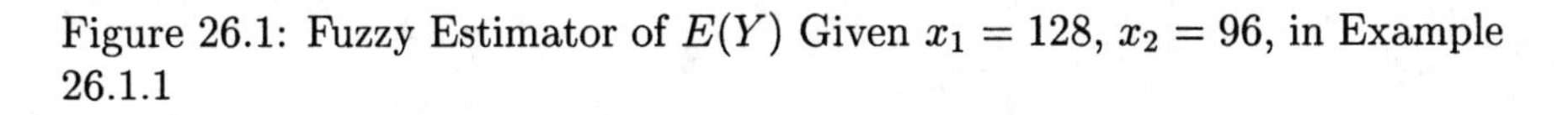

Figure 26.1: Fuzzy Estimator of $E(Y)$ Given $x_1 = 128$, $x_2 = 96$, in Example 26.1.1

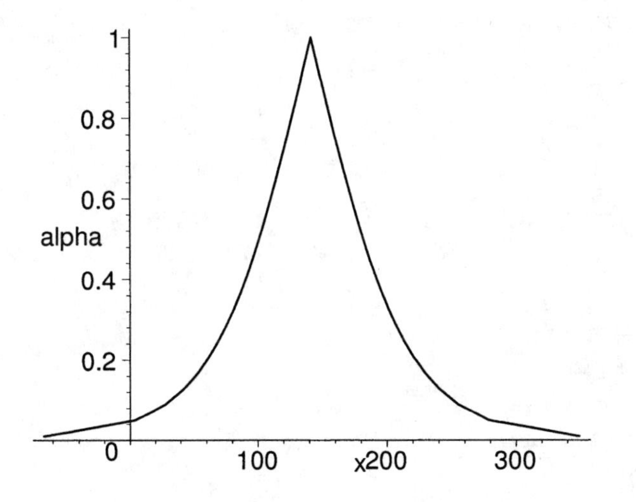

Figure 26.2: Fuzzy Estimator of $E(Y)$ Given $x_1 = 132$, $x_2 = 92$, in Example 26.1.1

Confidence Interval	$x_1 = 128, x_2 = 96$	$x_1 = 132, x_2 = 92$
$\overline{y}(x)[0]$	$[-71.98, 344.42]$	$[-66.97, 349.42]$
99% CI for $E(Y)$	$[126.38, 146.05]$	$[128.93, 153.49]$
99% CI for y	$[119.65, 152.77]$	$[123.09, 159.34]$

Table 26.1: Comparing the 99% Confidence Intervals in Example 26.1.1

26.2 References

1. J.Johnston: Econometric Methods, Second Edition, McGraw-Hill, N.Y., 1972.

2. Maple 6, Waterloo Maple Inc., Waterloo, Canada.

Chapter 27

Hypothesis Testing in Regression

27.1 Introduction

We look at two fuzzy hypothesis tests in this chapter: (1) in the next section $H_0 : b = 0$ verses $H_1 : b > 0$ a one-sided test; and (2) in the third section $H_0 : c = 0$ verses $H_1 : c \neq 0$ a two-sided test. In both cases we first review the crisp (non-fuzzy) test before the fuzzy test. We could also runs tests on a. However, we will continue to use the data in Table 25.1 were we determined $\widehat{a} = -49.3413$, so a is definitely negative and a test like $H_0 : a = 0$ verses $H_1 : a < 0$ seems a waste of time.

27.2 Tests on b

Let us first review the crisp situation. We wish to do the following hypothesis test

$$H_0 : b = 0, \tag{27.1}$$

verses

$$H_1 : b > 0, \tag{27.2}$$

which is a one-sided test. This is a one-sided test (see also Section 12.4 of Chapter 12, Example 19.2.1 of Chapter 19 and Section 24.2 of Chapter 24). Then we determine the statistic [1]

$$t_0 = \frac{\widehat{b} - 0}{\widehat{\sigma}\sqrt{a_{22}}}, \tag{27.3}$$

which, under H_0, has a t distribution with $(n-3)$ degrees of freedom. The a_{ii}, $1 \leq i \leq 3$, are the elements on the main diagonal of $(X^tX)^{-1}$ (see Section

25.2 of Chapter 25). Let γ, $0 < \gamma < 1$, be the significance level of the test. Usual values for γ are $0.10, 0.05, 0.01$. Our decision rule is: (1) reject H_0 if $t_0 \geq t_\gamma$; and (2) do not reject H_0 when $t_0 < t_\gamma$. In the above decision rule t_γ is the t-value so that the probability of a random variable, having the t probability density, exceeding t is γ. The critical region is $[t_\gamma, \infty)$ with critical value t_γ.

Now proceed to the fuzzy situation where our estimate of b is the triangular shaped fuzzy number $\overline{b}$ developed in Chapter 25. We will also need the fuzzy estimator for σ^2 also given in Chapter 25. Then our fuzzy statistic is

$$\overline{T} = \frac{\overline{b} - 0}{\overline{\sigma}\sqrt{a_{22}}}. \tag{27.4}$$

All fuzzy calculations will be performed via α-cuts and interval arithmetic. We find, after substituting the intervals for an alpha-cut of $\overline{b}$ and $\overline{\sigma}$ (square roots of equation (25.15)) into the expression for $\overline{T}$, using interval arithmetic, and simplification, that

$$\overline{T}[\alpha] = [\Pi_1(t_0 - t_{\alpha/2}), \Pi_2(t_0 + t_{\alpha/2})], \tag{27.5}$$

where

$$\Pi_1 = \sqrt{R(\lambda)/(n-3)}, \tag{27.6}$$

and

$$\Pi_2 = \sqrt{L(\lambda)/(n-3)}. \tag{27.7}$$

The $L(\lambda)$ and $R(\lambda)$ were defined in equations (25.13) and (25.14), respectively, in Chapter 25.

We have assumed that all intervals are positive in the derivation of equation (27.5). The interval for an alpha-cut of $\overline{b}$ may be positive or negative, but the interval for an alpha-cut of $\overline{\sigma}^2$ is always positive. When the left end point (or both end points) of the interval for an alpha-cut of $\overline{b}$ is negative we have to make some changes in equation (27.5). See section 13.3.1 in Chapter 13 for the details.

Now that we know the alpha-cuts of the fuzzy statistic we can find α-cuts of the fuzzy critical value $\overline{CV}_2$ for this one-sided test. As in previous chapters we get

$$\overline{CV}_2[\alpha] = [\Pi_1(t_\gamma - t_{\alpha/2}), \Pi_2(t_\gamma + t_{\alpha/2})]. \tag{27.8}$$

In this equation γ is fixed and alpha varies in the interval $[0.01, 1]$.

We now have a fuzzy set $\overline{T}$ for our test statistic and a fuzzy set $\overline{CV}_2$ for the critical value. Our final decision will depend on the relationship between $\overline{T}$ and $\overline{CV}_2$. Our test becomes : (1) reject H_0 if $\overline{T} > \overline{CV}_2$; (2) do not reject if $\overline{T} < \overline{CV}_2$; and (3) there is no decision on H_0 if $\overline{T} \approx \overline{CV}_2$.

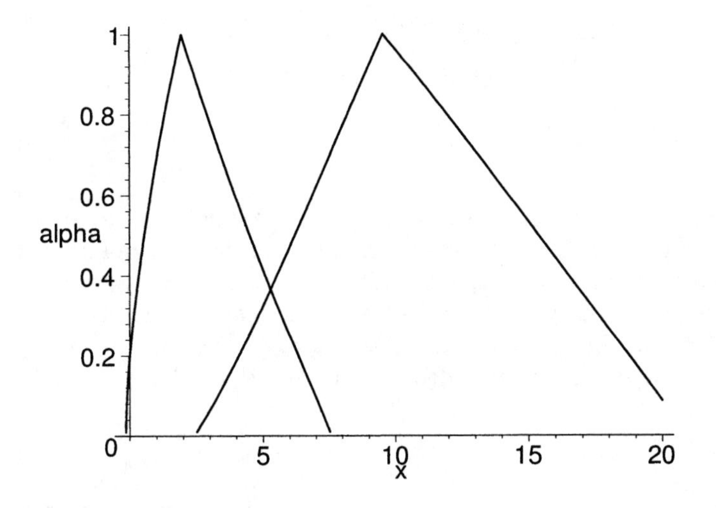

Figure 27.1: Fuzzy Test $\overline{T}$ verses $\overline{CV}_2$ in Example 27.2.1($\overline{CV}_2$ left, $\overline{T}$ right)

Example 27.2.1

We will still use the data in Table 25.1 and we have computed $\widehat{a} = -49.3413$, $\widehat{b} = 1.3642$. $\widehat{c} = 0.1139$ and $\widehat{\sigma}^2 = 12.9236$ with $n = 9$ and $a_{22} = 0.001586$. Let $\gamma = 0.05$, and determine $t_0 = 9.5287$ and $t_{0.05} = 1.943$ with 6 degrees of freedom. We compute

$$L(\lambda) = 18.548 - 12.548\lambda, \tag{27.9}$$

$$R(\lambda) = 0.676 + 5.324\lambda, \tag{27.10}$$

$$\Pi_1 = \sqrt{0.1127 + 0.8873\lambda}, \tag{27.11}$$

$$\Pi_2 = \sqrt{3.0913 - 2.0913\lambda}. \tag{27.12}$$

From these results we may get the graphs of $\overline{T}$ and $\overline{CV}_2$, using Maple [2], and they are shown in Figure 27.1.

From Figure 27.1 we see that $\overline{T} > \overline{CV}_2$. We therefore conclude: reject H_0. Of course, the crisp test would have the same result.

27.3 Tests on c

Let us first discuss the crisp hypothesis test. We wish to do the following hypothesis test

$$H_0 : c = 0, \tag{27.13}$$

verses

$$H_1 : c \neq 0, \tag{27.14}$$

which is a two-sided test. Next determine the statistic [1]

$$t_0 = \frac{\widehat{c} - 0}{\widehat{\sigma}\sqrt{a_{33}}}, \tag{27.15}$$

where, under H_0, t_0 has a t distribution with $(n-3)$ degrees of freedom. Let γ, $0 < \gamma < 1$, be the significance level of the test. Our decision rule is: (1) reject H_0 if $t_0 \geq t_{\gamma/2}$ or if $t_0 \leq -t_{\gamma/2}$; and (2) otherwise do not reject H_0.

Now proceed to the fuzzy situation where our estimate of c is the triangular shaped fuzzy number $\overline{c}$ and our fuzzy estimator $\overline{\sigma}^2$ of σ^2 is also a fuzzy number. These fuzzy estimators were deduced in Chapter 25. Then our fuzzy statistic is

$$\overline{T} = \frac{\overline{c} - 0}{\overline{\sigma}\sqrt{a_{33}}}. \tag{27.16}$$

All fuzzy calculations will be performed via α-cuts and interval arithmetic. We find, after substituting the intervals for an alpha-cuts of $\overline{c}$ and $\overline{\sigma}^2$ (square root of equation (25.15)) into the expression for $\overline{T}$, using interval arithmetic, assuming all intervals are positive, that

$$\overline{T}[\alpha] = [\Pi_1(t_0 - t_{\alpha/2}), \Pi_2(t_0 + t_{\alpha/2})], \tag{27.17}$$

where the Π_i were defined in the previous section.

Now that we know the alpha-cuts of the fuzzy statistic we can find α-cuts of the fuzzy critical values $\overline{CV}_i$, $i = 1, 2$. As in previous chapters we obtain

$$\overline{CV}_2[\alpha] = [\Pi_1(t_{\gamma/2} - t_{\alpha/2}), \Pi_2(t_{\gamma/2} + t_{\alpha/2})], \tag{27.18}$$

and $\overline{CV}_1 = -\overline{CV}_2$.

Given the fuzzy numbers $\overline{T}$ and the $\overline{CV}_i$, $i = 1, 2$, we may compare $\overline{T}$ to $\overline{CV}_1$ and then to $\overline{CV}_2$ to determine our final conclusion on H_0.

Example 27.3.1

We will still use the data in Table 25.1 and we have already computed $\widehat{c} = 0.1139$, $\widehat{\sigma}^2 = 12.9236$ with $n = 9$ and $a_{33} = 0.001591$. Let $\gamma = 0.05$, and compute $t_0 = 0.7943$ and $t_{0.025} = 2.447$ with 6 degrees of freedom.

The values of $L(\lambda)$, $R(\lambda)$, Π_1 and Π_2 are all the same as in Example 27.2.1. All that has changed is the value of t_0 and that now we use both $\overline{CV}_1$ and $\overline{CV}_2$ for a two-sided test.

The graphs of $\overline{T}$ and the $\overline{CV}_i$ are shown in Figure 27.2. It is evident that $\overline{CV}_2 > \overline{T}$, because the height of the intersection is less than 0.8. The point of intersection is close to 0.8, but slightly less than 0.8. Now compare $\overline{T}$ to $\overline{CV}_1$. The graph of $\overline{T}$ is not correct to the left of the vertical axis because in

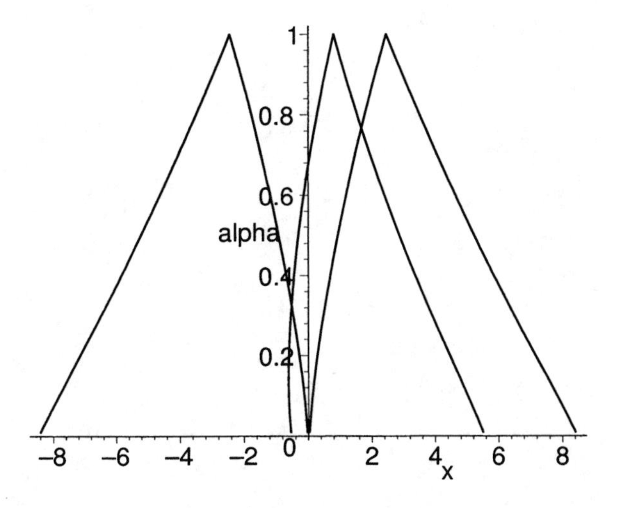

Figure 27.2: Fuzzy Test $\overline{T}$ verses the $\overline{CV}_i$ in Example 27.3.1 ($\overline{CV}_1$ left, $\overline{CV}_2$ right, $\overline{T}$ center)

computing the alpha-cuts of $\overline{T}$ we had assumed that all intervals were always positive, which is not true. However, there is no need to correct this because even if we did the height of the intersection of $\overline{T}$ and $\overline{CV}_1$ would be less than 0.8. You see that the left side of $\overline{T}$ crosses the vertical axis below 0.8 so it must, even with corrections for non-positive intervals, cross the right side of $\overline{CV}_1$ below 0.8 also. Hence $\overline{CV}_1 < \overline{T} < \overline{CV}_2$ and we do not reject H_0 which supports the hypothesis that $c = 0$. The same is true in the crisp case.

27.4 References

1. J.Johnston: Econometric Methods, Second Edition, McGraw-Hill, N.Y., 1972.

2. Maple 6, Waterloo Maple Inc., Waterloo, Canada.

Chapter 28

Summary and Questions

28.1 Summary

This book introduces basic (elementary) fuzzy statistics based on crisp (non-fuzzy) data. After the Introduction and an brief survey of fuzzy sets (Chapter 2) the book is in three parts: (1) fuzzy estimation in Chapters 3-11; (2) fuzzy hypothesis testing in Chapters 12-20; and (3) fuzzy linear regression in Chapters 21-27. The key to the development is fuzzy estimation.

Consider a normally distributed population with unknown mean and variance. We gather a random sample from this population in order to estimate its mean and variance. From the data we compute the sample mean $\overline{x}$ and variance s^2. Quite often we use these two point estimators in further calculations such as in hypothesis testing. However, these point estimators do not show their uncertainty so we could use a confidence interval for the unknown mean and variance. But which confidence interval? What we did was to use all $(1-\beta)100\%$ confidence intervals for $\xi \leq \beta \leq 1$. In this book we usually used $\xi = 0.01$ but one could have $\xi = 0.001$. We placed the confidence intervals one on top of another to obtain a triangular shaped fuzzy number as our fuzzy estimator of both the mean and variance. Once we have our fuzzy estimators the rest of the book follows.

In hypothesis testing one has a test statistic to be evaluated and compared to critical values to decide to reject, or not to reject, the null hypothesis. To evaluate the test statistic one usually needs estimates of population parameters like the mean and variance. But these estimators now become fuzzy numbers so the test statistic will also be a fuzzy number. It follows that the critical values will also be fuzzy numbers. We need to compare fuzzy numbers, for the test statistic and the critical values, do see if we will reject, or do not reject, the null hypothesis. This comprises the fuzzy hypothesis testing part of the book.

Next consider the basic simple linear regression equation $y = a + b(x - \overline{x})$.

We must estimate a and b from the crisp data which leads to fuzzy estimators. With fuzzy numbers for a and b we have fuzzy regression and fuzzy prediction. Fuzzy hypothesis testing on a and b is also contained in this topic. We consider both simple linear regression and multiple linear regression.

28.2 Questions

Certain decisions were made in the book which will now be formulated as "questions".

28.2.1 Unbiased Fuzzy Estimators

We faced the problem of getting an unbiased fuzzy estimator in Chapters 6, 11, 22 and 25. We said that our fuzzy estimator was unbiased if the vertex (where the membership function equals one) of the fuzzy number is at the crisp point estimator. For example, if we are estimating the variance of a normally distributed population, the vertex should be at s^2 the sample variance. Otherwise, it is a biased fuzzy estimator. Using the usual confidence intervals to construct our fuzzy estimator produced a biased fuzzy estimator. We proposed a solution to this problem giving an unbiased fuzzy estimator. Is there a "better" solution?

28.2.2 Comparing Fuzzy Numbers

In fuzzy hypothesis testing we need to be able to determine which of the following three possibilities for two fuzzy numbers $\overline{M}$ and $\overline{N}$ is true: (1) $\overline{M} < \overline{N}$, (2) $\overline{M} \approx \overline{N}$, or (3) $\overline{M} > \overline{N}$. In this book we used the procedure outlined in Section 2.5 in Chapter 2. If we use another method of comparing two fuzzy numbers (see the references in Chapter 2), how will the results of fuzzy hypothesis testing be effected? Is fuzzy hypothesis testing robust with respect to the method of comparing fuzzy numbers?

28.2.3 No Decision Conclusion

In Chapter 12 our final conclusion in fuzzy hypothesis testing was: (1) reject the null hypothesis; (2) do not reject the null hypothesis; or (3) no decision on the null hypothesis. Let the fuzzy test statistic be $\overline{Z}$ and the two fuzzy critical values $\overline{CV}_i$, $i = 1, 2$. The "no decision" conclusion resulted from $\overline{CV}_1 \approx \overline{Z} < \overline{CV}_2$ or $\overline{CV}_1 < \overline{Z} \approx \overline{CV}_2$. In this case should the final decision be "do not reject the null hypothesis"?

28.2.4 Equal Sample Sizes

In Chapters 17 and 20 we took a random sample of size n_1 (n_2) from population I (II). We ran into trouble finding a unique α-cut to relate to these sample sizes if they were not equal. So we took the easy way out by assuming that $n_1 = n_2$. We did suggest in Section 17.4 a compromise solution for α when the sample sizes are not equal. How shall we solve the problem of no unique α when $n_1 \neq n_2$?

28.2.5 Future

In the Introduction we mentioned that we cover most of elementary statistics that can be found in an introductory course in statistic except contingency tables, ANOVA and nonparametric statistics. So where to next? Our method starts with crisp data producing fuzzy number estimators. Any statistical method based on estimation, not covered in this book, could be next.

Chapter 29

Maple Commands

29.1 Introduction

In this chapter we give some of the Maple commands for the figures in the book. The commands are for Maple 6 but should be good for Maple 7, 8 and 9.

When you do a figure in Maple and then export it to LaTeX you get two files. The first one is in LaTeX and the second is a "eps" file. In the "eps" file go to the " (backslash)drawborder true def" line and change it to "(backslash)drawborder false def" if you do not want the border. Also for the figures we will use y in place of α (or λ) for the variable in the vertical axis.

29.2 Chapter 3

The Maple commands for Figure 3.2 are:

1. with(plots);
2. with(stats);
3. f1:=y->28.6-(1.25)*(statevalf[icdf,normald](1-(y/2)));
4. f2:=y->28.6+(1.25)*(statevalf[icdf,normald](1-(y/2)));
5. eq1:=x=f1(y);
6. eq2:=x=f2(y);
7. implicitplot({eq1,eq2},x=20..36,y=0.1..1,color=black,thickness=3, labels=[x,alpha]);
   (for the other graphs use $y = 0.01..1$ or $y = 0.001..1$)

29.3 Chapter 4

The Maple commands for Figure 4.1 are:

1. with(plots);
2. with(stats);
3. f1:=y->28.6-(0.3699)*(statevalf[icdf,studentst[24]](1-(y/2)));
4. f2:=y->28.6+(0.3699)*(statevalf[icdf,studentst[24]](1-(y/2)));
5. eq1:=x=f1(y);
6. eq2:=x=f2(y);
7. implicitplot({eq1,eq2},x=20..36,y=0.01..1,color=black,thickness=3, labels=[x,alpha]);
   (for the other graphs use $y = 0.10..1$ or $y = 0.001..1$)

29.4 Chapter 5

The Maple commands for Figure 5.1 are:

1. with(plots);
2. with(stats);
3. f1:=y->0.5143-(0.0267)*(statevalf[icdf,normald](1-(y/2)));
4. f2:=y->0.5143+(0.0267)*(statevalf[icdf,normald](1-(y/2)));
5. eq1:=x=f1(y);
6. eq2:=x=f2(y);
7. implicitplot({eq1,eq2},x=0..1,y=0.01..1,color=black,thickness=3, labels=[x,alpha]);
   (for the other graphs use $y = 0.10..1$ or $y = 0.001..1$)

29.5 Chapter 6

We substitute y for λ and then the Maple commands for Figure 6.1 are:

1. with(plots);
2. f1:=y->(1-y)*45.559 + y*24;
3. f2:=y->(1-y)*9.886 + y*24;

4. eq1:=x=82.08/f1(y);

5. eq2:=x=82.08/f2(y);

6. implicitplot({eq1,eq2},x=0..10,y=0.01..1,color=black,thickness=3, labels=[x,alpha]);
   (for the other graphs use $y = 0.10..1$ or $y = 0.001..1$)

29.6 Chapter 7

The Maple commands for Figure 7.1 are like those for Figure 3.2 and Figure 5.1.

29.7 Chapter 8

The Maple commands for Figures 8.1 and 8.2 are similar to those for Figure 4.1.

29.8 Chapter 9

Maple commands for Figure 9.1 are similar to those for Figure 4.1.

29.9 Chapter 10

Maple command for Figure 10.1 are the same as for Figure 3.2 in Chapter 3.

29.10 Chapter 11

The Maple commands for Figure 11.1 are (using y for λ):

1. with(plots);

2. f1:=y->((1-y)*0.1705 + y)*(1.459);

3. f2:=y->((1-y)*8.2697 + y)*(1.459);

4. eq1:=x=f1(y);

5. eq2:=x=f2(y);

6. implicitplot({eq1,eq2},x=0..20,y=0.01..1,color=black,thickness=3, labels=[x,alpha]);

29.11 Chapter 12

Maple commands for Figure 12.1 are:

1. with(plots);
2. with(stats);
3. f1:=y->1.6+(statevalf[icdf,normald](1-(y/2)));
4. f2:=y->1.6-(statevalf[icdf,normald](1-(y/2)));
5. f3:=y->1.96+(statevalf[icdf,normald](1-(y/2)));
6. f4:=y->1.96-(statevalf[icdf,normald](1-(y/2)));
7. eq1:=x=f1(y);
8. eq2:=x=f2(y);
9. eq3:=x=f3(y);
10. eq4:=x=f4(y);
11. implicitplot({eq1,eq2,eq3,eq4},x=-10..10,y=0.01..1,color=black, thickness=3,labels=[x,alpha]);

29.12 Chapter 13

We give the Maple commands for Figure 13.1. All the numerical values of the data items have been loaded into the functions. We use y for λ.

1. with(plots);
2. with(stats);
3. f1:=y-> $\sqrt{1.40169 - 0.40169 * y}$;
   ($f_1(y) = \Pi_2$)
4. f2:=y-> $\sqrt{0.67328 + 0.32672 * y}$;
   ($f_2(y) = \Pi_1$)
5. L:=y-> $140.169 - 40.169 * y$;
6. R:=y-> $67.328 + 32.672 * y$;
7. z2:=1-statevalf[cdf,chisquare[100]](L(y));
8. z1:=statevalf[cdf,chisquare[100]](R(y));
9. z:=z1+z2;
   (computes $\alpha = z = f(\lambda)$, $y = \lambda$)

10. eq1:=x=f2(y)*(1.6-statevalf[icdf,studentst[100]](1-(z/2)));
    (the left side of $\overline{T}$)

11. eq2:=x=f1(y)*(1.6+statevalf[icdf,studentst[100]](1-(z/2)));
    (the right side of $\overline{T}$)

12. eq3:=x=f2(y)*(2.626-statevalf[icdf,studentst[100]](1-(z/2)));
    (the left side of $\overline{CV}_2$)

13. eq3:=x=f1(y)*(2.626+statevalf[icdf,studentst[100]](1-(z/2)));
    (the right side of $\overline{CV}_2$)

14. implicitplot({eq1,eq2,eq3,eq4},x=-10..10,y=0.01..1,color=black,
    thickness=3,labels=[x,alpha]);

29.13 Chapter 14

Maple commands for Figure 14.1 are:

1. with(plots);
2. with(stats);
3. f1:=y->0.8-(0.9968)*(statevalf[icdf,normald](1-(y/2)));
4. f2:=y->0.8+(0.9968)*(statevalf[icdf,normald](1-(y/2)));
5. f3:=y->1.96-(0.9968)*(statevalf[icdf,normald](1-(y/2)));
6. f4:=y->1.96-(0.9968)*(statevalf[icdf,normald](1-(y/2)));
7. eq1:=x=f1(y);
8. eq2:=x=f2(y);
9. eq3:=x=f3(y);
10. eq4:=x=f4(y);
11. implicitplot({eq1,eq2,eq3,eq4},x=-5..5,y=0.01..1,color=black,
    thickness=3,labels=[x,alpha]);

29.14 Chapter 15

The Maple commands for Figure 15.1 are, using y for λ:

1. with(plots);
2. L:=y->140.169 -40.169*y;

3. R:=y->67.328+32.672*y;

4. eq1:=x=8375/L(y);
   (left side $\overline{\chi}^2$)

5. eq2:=x=8375/R(y);
   (right side $\overline{\chi}^2$)

6. eq3:=x=6732.8/L(y);
   (left side $\overline{CV}_1$)

7. eq4:=x=6732.8/R(y);
   (right side $\overline{CV}_1$)

8. eq5:=x=100*140.169/L(y);
   (left side $\overline{CV}_2$)

9. eq6:=x=100*140.169/R(y);
   (right side $\overline{CV}_2$)

10. implicitplot({eq1,eq2,eq3,eq4,eq5,eq6 },x=0..300,y=0.01..1,color=black, thickness=3,labels=[x,alpha]);

29.15 Chapter 16

The Maple commands for Figure 16.1 are:

1. with(plots);

2. with(stats);

3. f1:=y->-5.48-(statevalf[icdf,normald](1-(y/2)));

4. f2:=y->-5.48+(statevalf[icdf,normald](1-(y/2)));

5. f3:=y->-1.96-(statevalf[icdf,normald](1-(y/2)));

6. f4:=y->-1.96+(statevalf[icdf,normald](1-(y/2)));

7. eq1:=x=f1(y);

8. eq2:=x=f2(y);

9. eq3:=x=f3(y);

10. eq4:=x=f4(y);

11. implicitplot({eq1,eq2,eq3,eq4},x=-10..10,y=0.01..1,color=black, thickness=3,labels=[x,alpha]);

29.16 Chapter 17

We give the Maple commands for Figure 17.1. All the numerical values of the data items have been loaded into the functions. We use y for λ.

1. with(plots);
2. with(stats);
3. f1:=y-> $\sqrt{0.517675 + 0.482325 * y}$;
   ($f_1(y) = \Pi_1$)
4. f2:=y-> $\sqrt{1.66915 - 0.66915 * y}$;
   ($f_2(y) = \Pi_2$)
5. L:=y-> $66.766 - 26.766 * y$;
6. R:=y-> $20.707 + 19.293 * y$;
7. z2:=1-statevalf[cdf,chisquare[40]](L(y));
8. z1:=statevalf[cdf,chisquare[40]](R(y));
9. z:=z1+z2;
   (computes $\alpha = z = f(\lambda)$, $y = \lambda$)
10. eq1:=x=f1(y)*(-1.748+statevalf[icdf,normald](1-(z/2)));
    (the right side of $\overline{Z}$)
11. eq2:=x=f2(y)*(-1.748-statevalf[icdf,normald](1-(z/2)));
    (the left side of $\overline{Z}$)
12. eq3:=x=f1(y)*(-1.96+statevalf[icdf,normald](1-(z/2)));
    (the right side of $\overline{CV}_1$)
13. eq4:=x=f2(y)*(-1.96-statevalf[icdf,normald](1-(z/2)));
    (the left side of $\overline{CV}_1$)
14. implicitplot({eq1,eq2,eq3,eq4},x=-10..10,y=0.01..1,color=black, thickness=3,labels=[x,alpha]);

The Maple commands for Figure 17.3 are:

1. with(plots);
2. with(stats);
3. f1:=y-> $\sqrt{0.3717 + 0.6283 * y}$;
   ($f_1(y) = \Pi_1$)
4. f2:=y-> $\sqrt{1.9998 - 0.9998 * y}$;
   ($f_2(y) = \Pi_2$)

5. L:=y-> $39.997 - 19.997 * y$;
6. R:=y-> $7.434 + 12.566 * y$;
7. z2:=1-statevalf[cdf,chisquare[20]](L(y));
8. z1:=statevalf[cdf,chisquare[20]](R(y));
9. z:=z1+z2;
   (computes $\alpha = z = f(\lambda)$, $y = \lambda$)
10. eq1:=x=f2(y)*(1.5905+statevalf[icdf,studentst[20]](1-(z/2)));
   (the right side of $\overline{T}$)
11. eq2:=x=f1(y)*(1.5905-statevalf[icdf,studentst[20]](1-(z/2)));
   (the left side of $\overline{T}$)
12. eq3:=x=f2(y)*(2.022+statevalf[icdf,studentst[20]](1-(z/2)));
   (the right side of $\overline{CV}_2$)
13. eq4:=x=f2(y)*(2.022-statevalf[icdf,studentst[20]](1-(z/2)));
   (the left side of $\overline{CV}_2$)
14. implicitplot({eq1,eq2,eq3,eq4},x=-1..10,y=0.01..1,color=black, thickness=3,labels=[x,alpha]);

29.17 Chapter 18

The Maple instructions for Figure 18.1 are:

1. with(plots);
2. with(stats);
3. f1:=y->-1.645-(0.9293)*(statevalf[icdf,normald](1-(y/2)));
4. f2:=y->-1.645+(0.9293)*(statevalf[icdf,normald](1-(y/2)));
5. f3:=y->-1.96-(0.99293)*(statevalf[icdf,normald](1-(y/2)));
6. f4:=y->-1.96+(0.9293)*(statevalf[icdf,normald](1-(y/2)));
7. eq1:=x=f1(y);
8. eq2:=x=f2(y);
9. eq3:=x=f3(y);
10. eq4:=x=f4(y);
11. implicitplot({eq1,eq2,eq3,eq4},x=-10..10,y=0.01..1,color=black, thickness=3,labels=[x,alpha]);

29.18 Chapter 19

Maple commands for Figure 19.1 are:

1. with(plots);
2. with(stats);
3. f1:=y-> $\sqrt{2.621 - 1.621 * y}$;
   $(f_1(y) = \Pi_2)$
4. f2:=y-> $\sqrt{0.193 + 0.807 * y}$;
   $(f_2(y) = \Pi_1)$
5. L:=y-> $23.589 - 14.589 * y$;
6. R:=y-> $1.735 + 7.265 * y$;
7. z2:=1-statevalf[cdf,chisquare[9]](L(y));
8. z1:=statevalf[cdf,chisquare[9]](R(y));
9. z:=z1+z2;
   (computes $\alpha = z = f(\lambda)$, $y = \lambda$)
10. eq1:=x=f2(y)*(-1.2104+statevalf[icdf,studentst[9]](1-(z/2)));
    (the right side of $\overline{T}$)
11. eq2:=x=f1(y)*(-1.2104-statevalf[icdf,studentst[9]](1-(z/2)));
    (the left side of $\overline{T}$)
12. eq3:=x=f2(y)*(-2.821+statevalf[icdf,studentst[9]](1-(z/2)));
    (the right side of $\overline{CV}_1$)
13. eq4:=x=f1(y)*(-2.821-statevalf[icdf,studentst[9]](1-(z/2)));
    (the left side of $\overline{CV}_1$)
14. implicitplot({eq1,eq2,eq3,eq4},x=-10..10,y=0.01..1,color=black, thickness=3,labels=[x,alpha]);

29.19 Chapter 20

We give the Maple commands for Figure 20.1. We use y in place of λ. Also, all the numerical values from the example have been loaded into the functions.

1. with(plots);
2. f11:=y->25.188-15.188*y;
3. f12:=y->2.156+7.844*y;

4. f21:=y->28.299-16.299*y;

5. f22:=y->3.074+8.926*y;

6. f1:=->(10*f22(y))/(12*f11(y));
(this is Γ_1)

7. f2:=y->((10*f21(y))/(12*f12(y));
(this is Γ_2)

8. eq1:=x=f1(y)*(0.24/0.35);
(left side $\overline{F}$)

9. eq2:=x=f2(y)*(0.24/0.35);;
(right side $\overline{F}$)

10. eq3:=x=f1(y)*(3.3736);
(left side $\overline{CV}_2$)

11. eq4:=x=f2(y)*(3.3736);
(right side $\overline{CV}_2$)

12. eq5:=x=f1(y)*(0.2762);
(left side $\overline{CV}_1$)

13. eq6:=x=f2(y)*(0.2762);
(right side $\overline{CV}_1$)

14. implicitplot({eq1,eq2,eq3,eq4,eq5,eq6 },x=0..20,y=0..1,color=black, thickness=3,labels=[x,lambda]);

29.20 Chapter 21

The Maple commands for Figure 21.1 are:

1. with(plots);

2. with(stats);

3. f1:=y->statevalf[icdf,normald](1-(y/2));

4. f2:=y->1.35-0.65*exp((2/sqrt(13))*f1(y));

5. f3:=y->1.35+0.65*exp((2/sqrt(13))*f1(y));

6. f4:=y->1.35-0.65*exp((-2/sqrt(13))*f1(y));

7. f5:=y->1.35+0.65*exp((-2/sqrt(13))*f1(y));

8. eq1:=x=f2(y)/f3(y);

9. eq2:=x=f4(y)/f5(y);
10. implicitplot({eq1,eq2},x=-1..1,y=0.01..1,color=black,thickness=3, labels=[x,alpha]);

29.21 Chapter 22

The Maple commands for Figures 22.1 and 22.2 are similar to those in Chapter 4 and are omitted.

The Maple commands for Figure 22.3 are (using y for λ):

1. with(plots);
2. f1:=y->(1-y)*21.955 + y*10;
3. f2:=y->(1-y)*1.344 + y*10;
4. eq1:=x=217.709/f1(y);
5. eq2:=x=217.709/f2(y);
6. implicitplot({eq1,eq2},x=0..50,y=0.01..1,color=black,thickness=3, labels=[x,alpha]);

29.22 Chapter 23

The Maple commands for Figure 23.1 are:

1. with(plots);
2. with(stats);
3. f1:=y->81.3-(1.6496)*(statevalf[icdf,studentst[8]](y/2));
4. f2:=y->81.3+(1.6496)*(statevalf[icdf,studentst[8]](y/2)); (two sides of $\overline{a}$)
5. f3:=y->0.742+(0.1897)*(statevalf[icdf,studentst[8]](y/2));
6. f4:=y->0.742-(0.1897)*(statevalf[icdf,studentst[8]](y/2)); (two sides of $\overline{b}$)
7. g1:=y-> f1(y)+(-8.3)*f4(y);
8. g2:=y-> f2(y)+(-8.3)*f3(y);
9. eq1:=x=g1(y);
10. eq2:=x=g2(y);
11. implicitplot({eq1,eq2},x=60..90,y=0.01..1,color=black,thickness=3, labels=[x,alpha]);

29.23 Chapter 24

The Maple commands for Figure 24.1 are (using y for λ):

1. with(plots);

2. with(stats);

3. f2:=y-> $\sqrt{2.1955 - 1.1955 * y}$;
   ($f_2(y) = \Pi_2$)

4. f1:=y-> $\sqrt{0.1344 + 0.8656 * y}$;
   ($f_1(y) = \Pi_1$)

5. L:=y-> $21.955 - 11.955 * y$;

6. R:=y-> $1.344 + 8.656 * y$;

7. z2:=1-statevalf[cdf,chisquare[8]](L(y));

8. z1:=statevalf[cdf,chisquare[8]](R(y));

9. z:=z1+z2;
   (computes $\alpha = z = f(\lambda)$, $y = \lambda$)

10. eq1:=x=f2(y)*(0.7880+statevalf[icdf,studentst[8]](1-(z/2)));
    (the right side of $\overline{T}$)

11. eq2:=x=f1(y)*(0.7880-statevalf[icdf,studentst[8]](1-(z/2)));
    (the left side of $\overline{T}$)

12. eq3:=x=f2(y)*(1.860+statevalf[icdf,studentst[8]](1-(z/2)));
    (the right side of $\overline{CV}_2$)

13. eq3:=x=f1(y)*(1.860-statevalf[icdf,studentst[8]](1-(z/2)));
    (the left side of $\overline{CV}_2$)

14. implicitplot({eq1,eq2,eq3,eq4},x=-10..10,y=0.01..1,color=black, thickness=3,labels=[x,alpha]);

29.24 Chapter 25

The Maple commands for Figures 25.1-25.3 are similar to those for Figure 4.1. The Maple commands for Figure 25.4 are like those for Figure 6.1.

29.25 Chapter 26

The Maple commands for Figure 26.1 are:

1. with(plots);
2. with(stats);
3. f11:=y->-49.3413-(24.0609)*(statevalf[icdf,studentst[6]](1-(y/2)));
4. f12:=y->-49.3413+(24.0609)*(statevalf[icdf,studentst[6]](1-(y/2)));
5. f21:=y->128*(1.3642-(0.1432)*(statevalf[icdf,studentst[6]](1-(y/2)));
6. f22:=y->128*(1.3642+(0.1432)*(statevalf[icdf,studentst[6]](1-(y/2)));
7. f31:=y->96*(0.1139-(0.1434)*(statevalf[icdf,studentst[6]](1-(y/2)));
8. f32:=y->96*(0.1139+(0.1434)*(statevalf[icdf,studentst[6]](1-(y/2)));
9. eq1:=x=f11(y)+f21(y)+f31(y);
10. eq2:=x=f12(y)+f22(y)+f32(y);
11. implicitplot({eq1,eq2},x=-100..400,y=0.01..1,color=black,thickness=3, labels=[x,alpha]);

29.26 Chapter 27

The Maple commands for Figures 27.1 and 27.2 are similar to those for Figure 24.1.

Index

alpha
  function of lambda, 33, 63, 65, 74, 83, 97, 100, 109, 125, 141
alpha-cut, 2, 3, 5, 6, 10, 17, 18, 54, 104, 113, 129, 134, 136
  function of lambda, 100
alpha-cuts, 62, 78

beta
  function of lambda, 33, 51, 109, 125
binomial distribution, 3, 27, 47, 69, 91
bivariate normal distribution, 103

chi-square distribution, 31, 32, 55, 73, 82, 99, 124
  critical values, 3
  mean, 33
  median, 33
compromise solution, 89, 100, 141
confidence interval, 140
  $E(Y)$
    new value x, 114
    new values x, 130
  β, 2, 17
  μ, 19, 23, 24
  $\mu_1 - \mu_2$
    large samples, 39
    variances equal, 40
    variances known, 37
    variances unequal, 41
    variances unknown, 39
  μ_d, 43
  ρ, 104
  σ, 33
  σ^2, 32, 34, 108, 109, 124
  σ_1^2/σ_2^2, 49, 51
  σ_1/σ_2, 50, 51
  θ, 17
  a, 108, 109, 114, 123, 124, 130
  b, 108, 109, 114, 123, 124, 130
  c, 123, 124, 130
  p, 27, 28
  $p_1 - p_2$, 48
  y
    new value x, 114
    new values x , 130
  $y(x)$
    new value x , 114
  $y(x_1, x_2)$
    new values x, 130
  mean, 139
  variance, 139
crisp critical region, 117, 134
crisp critical values, 54, 61, 78, 96, 139
crisp data, 1, 107, 123, 139, 140
crisp decision rule, 54, 61, 69, 73, 78, 82, 91, 95, 100, 104, 117, 120, 134, 136, 139
crisp estimator, 139
  σ^2
    unbiased, 23
  p
    pooled, 92
  confidence interval, 18
  point, 18, 23, 139, 140
    $\overline{x}$, 19
    ρ, 103
    σ^2, 107, 124

θ, 17
a, 107, 123
a,b,c, 124
b, 107, 123
c, 123
p, 27, 69
$p_1 - p_2$, 47
p_i, 47, 91
s^2, 31
unbiased, 31
pooled
variance, 40
crisp function, 5
crisp hypothesis test, 139
μ
variance known, 53
variance unknown, 61
$\mu_1 = \mu_2$
equal sample sizes, 86, 87
large samples, 81
small samples, 81, 86, 87
variances equal, 81, 86
variances known, 77
variances unequal, 81, 87
variances unknown, 81
$\mu_d = 0$, 95
ρ, 103
σ^2, 73
$\sigma_1^2 = \sigma_2^2$, 99
a, 117
b, 119, 133
c, 135
p, 69
$p_1 = p_2$, 91
crisp linear regression, 107, 139
crisp multiple regression, 123
crisp number, 2, 5
crisp set, 2, 5
crisp solution, 5
crisp statistics, 2, 3, 114, 130
ANOVA, 2, 141
contingency tables, 2, 141
nonparametric, 2, 141
crisp subset, 5
crisp test statistic, 1, 55, 139
μ
variance known, 54
variance unknown, 61
$\mu_1 = \mu_2$
large samples, 81
small samples, 86, 87
variances equal, 86
variances known, 77
variances unequal, 87
variances unknown, 81
$\mu_d = 0$, 95
ρ, 104
σ^2, 73
$\sigma_1^2 = \sigma_2^2$, 99
a, 117
b, 120, 133
c, 136
p, 69
$p_1 = p_2$, 91, 92
matched pairs, 43

degrees of freedom, 3, 87, 90
dependent samples, 43, 95

eps file, 143
extension principle, 2, 5, 8, 10

F distribution, 49–51, 55, 100
critical values, 3
factor, 32
failure, 27
probability, 27
figures, 3
LaTeX, 3, 143
Maple, 3, 19, 24, 28, 34, 38, 40, 42, 44, 48, 51, 56, 65, 70, 97, 110, 114, 135, 143
forecast high temperatures, 44
fuzzy arithmetic, 8
addition, 8
alpha-cuts and interval arithmetic, 2, 10, 54, 70, 74, 78, 82, 89, 96, 100, 104,

113, 118, 120, 129, 134, 136
division, 8
multiplication, 8
subtraction, 8
fuzzy correlation, 1, 103
fuzzy critical values, 54, 62, 70, 74, 78, 84, 87, 89, 92, 96, 100, 105, 118, 120, 134, 136, 139
fuzzy data, 1, 107, 123, 139
fuzzy decision rule, 56, 59, 63, 65, 70, 74, 78, 139
do not reject, 56, 66, 71, 74, 97, 102, 105, 119, 137
no decision, 56, 57, 65, 75, 85, 88, 93, 101, 140
reject, 56, 57, 71, 78, 86, 121, 135
fuzzy estimator, 1–3, 8, 17, 18, 53, 139
$E(Y)$, 113
$\overline{\mu}$, 1, 17, 19, 24, 54, 62
$\overline{\mu}_d$, 44, 96
$\overline{\mu}_{12}$, 39, 78, 82, 87, 89
large samples, 39
small samples, 39, 41
variances equal, 39, 40
variances known, 37
variances unequal, 41
variances unknown, 39
$\overline{\rho}$, 104
$\overline{\sigma}$, 33, 34
$\overline{\sigma}^2$, 31–33, 62, 74, 82, 89, 96, 100, 108, 110, 118, 120, 125, 134, 136
$\overline{\sigma}^2_{12}$, 49–51
$\overline{\sigma}_{12}$, 51
$\overline{a}$, 108, 110, 113, 118, 124, 125, 129, 140
$\overline{b}$, 108, 110, 113, 120, 124, 125, 129, 134, 140
$\overline{c}$, 124, 125, 129, 136
$\overline{p}$, 28, 70
$\overline{p}_{12}$, 47, 48, 92
$\overline{y}(x_1, x_2)$, 129
biased, 31, 32, 49, 50, 108, 125, 140
unbiased, 31–33, 49–51, 108, 125, 140
fuzzy function, 11–13
alpha-cuts and interval arithmetic, 11–13
extension principle, 11, 13
membership function, 12
fuzzy functions, 5
fuzzy hypothesis testing, 1, 2, 5, 14, 139, 140
μ
variance known, 53
variance unknown, 62
$\mu_1 = \mu_2$
equal sample sizes, 83, 87
large samples, 82
small samples, 87
unequal sample sizes, 88
variances equal, 87
variances unequal, 87
$\mu_d = 0$, 96
ρ, 105
σ^2, 74
$\sigma_1^2 = \sigma_2^2$, 100
a, 117, 133, 140
b, 117, 133, 140
c, 133
p, 70
$p_1 = p_2$, 92
equal sample sizes, 141
fuzzy multiple regression, 1, 123, 129
fuzzy number, 1, 5, 6, 8, 9, 11, 12, 17
$\geq$ constant, 8
$\leq$, 14
base, 7
core, 7, 15
from confidence intervals, 17
membership function, 18

normalized, 18
support, 7
triangular, 6, 12, 51
triangular shaped, 3, 6, 18, 19, 28, 54, 55, 63, 70, 74, 78, 108, 118, 120, 134, 136, 139
fuzzy numbers, 2, 3, 5, 14, 113
<, 8
height of intersection, 15, 56, 57, 65, 70, 71, 74, 78, 85, 86, 88, 93, 97, 101, 105, 119, 121, 136, 137
partitioning, 14
fuzzy prediction, 1, 2, 113, 129, 140
fuzzy regression, 1, 2, 107, 113, 139, 140
fuzzy set
discrete, 8
membership function, 5
subset, 2, 5, 8
fuzzy sets, 1, 5, 139
$\leq$, 8
fuzzy statistics, 139
fuzzy test statistic, 139
μ
variance known, 54
variance unknown, 62
$\mu_1 = \mu_2$
equal sample sizes, 82, 83, 87
large samples, 82, 83
small samples, 87
unequal sample sizes, 88, 89
variances equal, 87
variances known, 78
variances unequal, 87
variances unknown, 82, 83
$\mu_d = 0$, 96
ρ, 104
σ^2, 74
$\sigma_1^2 = \sigma_2^2$, 100
a, 118
b, 120, 134
c, 136
p, 70
$p_1 = p_2$, 92

high blood pressure, 43, 95

inf, 9
greatest lower bound, 9
interval arithmetic, 5, 9, 12, 13, 113, 129
addition, 9
division, 9
multiplication, 9
non-positive intervals, 63
positive intervals, 62
subtraction, 9

lambda-cut, 100
linear correlation coefficient, 103

Maple, 83
commands, 3, 20, 24, 28, 34, 51, 56, 65, 70, 74, 78, 85, 88, 93, 97, 101, 104, 110, 114, 119, 130, 143
commands Fig. 11.1, 145
commands Fig. 12.1, 146
commands Fig. 13.1, 146
commands Fig. 14.1, 147
commands Fig. 15.1, 147
commands Fig. 16.1, 148
commands Fig. 17.1, 149
commands Fig. 17.3, 149
commands Fig. 18.1, 150
commands Fig. 19.1, 151
commands Fig. 20.1, 151
commands Fig. 21.1, 152
commands Fig. 22.3, 153
commands Fig. 23.1, 153
commands Fig. 24.1, 154
commands Fig. 26.1, 155
commands Fig. 3.2, 143
commands Fig. 4.1, 144
commands Fig. 5.1, 144

commands Fig. 6.1, 144
implicitplot, 3, 20
point estimators, 125
matched pairs, 43, 95
matrix
X, 123
matrix notation, 123

normal distribution, 1, 18, 23, 24, 27, 31, 34, 37, 39, 43, 47, 49, 53, 55, 61, 69, 73, 74, 77, 81, 82, 91, 95, 99, 104, 107, 139
critical values, 2
notation
fuzzy set, 2, 5

one-sided test, 53, 59, 96, 117, 133
ordering
fuzzy numbers, 2, 5, 14, 56, 139, 140
transitive, 14

probability density function, 17
probability mass function, 17

random sample, 1, 17, 19, 23, 24, 27, 31, 34, 37, 47, 49, 51, 53, 91, 95, 96, 101, 103
$\overline{d}$, 43
$\widehat{p}$, 69
s_d^2, 43
mean, 2, 19, 23, 24, 37, 54, 56, 61, 65, 77, 81, 95, 99, 139
variance, 23, 24, 39, 49, 61, 65, 73, 81, 95, 99, 139
ranking
fuzzy numbers, 2, 5, 14, 56, 139, 140
transitive, 14

sample correlation coefficient, 103
SAT scores, 96
significance level
γ, 2, 54, 56, 61, 69, 73, 77, 82, 91, 95, 100, 104, 117, 120, 134, 136
significant linear correlation, 106
success, 27
probability, 27, 47, 69, 91
sup, 9
least upper bound, 9

t distribution, 23, 40, 43, 55, 61, 74, 87, 90, 95, 108, 117, 120, 124, 133, 136
critical values, 3
two-sided test, 53

List of Figures

2.1 Triangular Fuzzy Number $\overline{N}$ 6
2.2 Triangular Shaped Fuzzy Number $\overline{P}$ 7
2.3 The Fuzzy Number $\overline{C} = \overline{A} \cdot \overline{B}$ 11
2.4 Determining $\overline{M} < \overline{N}$ 15

3.1 Fuzzy Estimator $\overline{\mu}$ in Example 3.3.1, $0.01 \leq \beta \leq 1$ 20
3.2 Fuzzy Estimator $\overline{\mu}$ in Example 3.3.1, $0.10 \leq \beta \leq 1$ 21
3.3 Fuzzy Estimator $\overline{\mu}$ in Example 3.3.1, $0.001 \leq \beta \leq 1$ 21

4.1 Fuzzy Estimator $\overline{\mu}$ in Example 4.1.1, $0.01 \leq \beta \leq 1$ 24
4.2 Fuzzy Estimator $\overline{\mu}$ in Example 4.1.1, $0.10 \leq \beta \leq 1$ 25
4.3 Fuzzy Estimator $\overline{\mu}$ in Example 4.1.1, $0.001 \leq \beta \leq 1$ 25

5.1 Fuzzy Estimator $\overline{p}$ in Example 5.1.1, $0.01 \leq \beta \leq 1$ 28
5.2 Fuzzy Estimator $\overline{p}$ in Example 5.1.1, $0.10 \leq \beta \leq 1$ 29
5.3 Fuzzy Estimator $\overline{p}$ in Example 5.1.1, $0.001 \leq \beta \leq 1$ 29

6.1 Fuzzy Estimator $\overline{\sigma}^2$ in Example 6.3.1, $0.01 \leq \beta \leq 1$ 35
6.2 Fuzzy Estimator $\overline{\sigma}^2$ in Example 6.3.1, $0.10 \leq \beta \leq 1$ 35
6.3 Fuzzy Estimator $\overline{\sigma}^2$ in Example 6.3.1, $0.001 \leq \beta \leq 1$ 36
6.4 Fuzzy Estimator $\overline{\sigma}$ in Example 6.3.1, $0.01 \leq \beta \leq 1$ 36

7.1 Fuzzy Estimator $\overline{\mu}_{12}$ in Example 7.1.1, $0.01 \leq \beta \leq 1$ 38

8.1 Fuzzy Estimator $\overline{\mu}_{12}$ in Example 8.3.1.1, $0.01 \leq \beta \leq 1$ 41
8.2 Fuzzy Estimator $\overline{\mu}_{12}$ in Example 8.3.2.1, $0.01 \leq \beta \leq 1$ 42

9.1 Fuzzy Estimator $\overline{\mu}_d$ in Example 9.1.1, $0.01 \leq \beta \leq 1$ 45

10.1 Fuzzy Estimator $\overline{p}_{12}$ in Example 10.1.1, $0.01 \leq \beta \leq 1$ 48

11.1 Fuzzy Estimator $\overline{\sigma}^2_{12}$ of σ_1^2/σ_2^2 in Example 11.3.1, $0.01 \leq \beta \leq 1$ 52
11.2 Fuzzy Estimator $\overline{\sigma}_{12}$ of σ_1/σ_2 in Example 11.3.1, $0.01 \leq \beta \leq 1$ 52

12.1 Fuzzy Test $\overline{Z}$ verses $\overline{CV}_2$ in Example 12.3.1($\overline{Z}$ left, $\overline{CV}_2$ right) 57

12.2 Fuzzy Test $\overline{Z}$ verses $\overline{CV}_1$ in Example 12.3.1($\overline{CV}_1$ left, $\overline{Z}$ right) 58
12.3 Fuzzy Test $\overline{Z}$ verses $\overline{CV}_1$ in Example 12.3.2($\overline{Z}$ left, $\overline{CV}_1$ right) 58

13.1 Fuzzy Test $\overline{T}$ verses $\overline{CV}_2$ in Example 13.3.1($\overline{T}$ left, $\overline{CV}_2$ right) 66
13.2 Fuzzy Test $\overline{T}$ verses $\overline{CV}_1$ in Example 13.3.2($\overline{T}$ right, $\overline{CV}_1$ left) 67

14.1 Fuzzy Test $\overline{Z}$ verses $\overline{CV}_2$ in Example 14.3.1($\overline{Z}$ left, $\overline{CV}_2$ right) 71
14.2 Fuzzy Test $\overline{Z}$ verses $\overline{CV}_2$ in Example 14.3.2($\overline{Z}$ right, $\overline{CV}_2$ left) 72

15.1 Fuzzy Test in Example 15.3.1($\overline{CV}_1$ left, $\overline{\chi}^2$ middle, $\overline{CV}_2$ right) 75
15.2 Fuzzy Test in Example 15.3.2($\overline{CV}_1$ left, $\overline{\chi}^2$ middle, $\overline{CV}_2$ right) 76

16.1 Fuzzy Test $\overline{Z}$ verses $\overline{CV}_1$ in Example 16.3.1($\overline{Z}$ left, $\overline{CV}_1$ right) 79

17.1 Fuzzy Test $\overline{Z}$ verses $\overline{CV}_1$ in Example 17.2.1($\overline{Z}$ right,$\overline{CV}_1$ left) 85
17.2 Fuzzy Test $\overline{Z}$ verses $\overline{CV}_2$ in Example 17.2.2($\overline{Z}$ right,$\overline{CV}_2$ left) 86
17.3 Fuzzy Test $\overline{T}$ verses $\overline{CV}_2$ in Example 17.3.2.1($\overline{T}$ left,$\overline{CV}_2$ right) 88

18.1 Fuzzy Test $\overline{Z}$ verses $\overline{CV}_1$ in Example 18.2.1($\overline{Z}$ right,$\overline{CV}_1$ left) 93

19.1 Fuzzy Test $\overline{T}$ verses $\overline{CV}_1$ in Example 19.2.1($\overline{T}$ right, $\overline{CV}_1$ left) 97

20.1 Fuzzy Test $\overline{F}$ verses $\overline{CV}_1$ and $\overline{CV}_2$ in Example 20.2.1($\overline{F}$ center, $\overline{CV}_1$ left, $\overline{CV}_2$ right) 101
20.2 Fuzzy Test $\overline{F}$ verses $\overline{CV}_1$ and $\overline{CV}_2$ in Example 20.2.2($\overline{F}$ center, $\overline{CV}_1$ left, $\overline{CV}_2$ right) 102

21.1 Fuzzy Estimator $\overline{p}$ in Example 21.3.1 105
21.2 Fuzzy Test $\overline{Z}$ verses the $\overline{CV}_i$ in Example 21.3.2($\overline{CV}_1$ left, $\overline{Z}$ middle, $\overline{CV}_2$ right) 106

22.1 Fuzzy Estimator for a in Example 22.2.1 110
22.2 Fuzzy Estimator for b in Example 22.2.1 111
22.3 Fuzzy Estimator for σ^2 in Example 22.2.1 111

23.1 Fuzzy Estimator of $E(Y)$ given $x = 60$ in Example 23.1 . . . 114
23.2 Fuzzy Estimator of $E(Y)$ given $x = 70$ in Example 23.1 . . . 115

24.1 Fuzzy Test $\overline{T}$ verses $\overline{CV}_2$ in Example 24.2.1($\overline{T}$ left, $\overline{CV}_2$ right) 119
24.2 Fuzzy Test $\overline{T}$ verses the $\overline{CV}_i$ in Example 24.3.1 ($\overline{CV}_1$ left, $\overline{CV}_2$ center, $\overline{T}$ right) 121

25.1 Fuzzy Estimator $\overline{a}$ for a in Example 25.2.1 127
25.2 Fuzzy Estimator $\overline{b}$ for b in Example 25.2.1 127
25.3 Fuzzy Estimator $\overline{c}$ for c in Example 25.2.1 128
25.4 Fuzzy Estimator $\overline{\sigma}^2$ for σ^2 in Example 25.2.1 128

26.1 Fuzzy Estimator of $E(Y)$ Given $x_1 = 128$, $x_2 = 96$, in Example 26.1.1 131
26.2 Fuzzy Estimator of $E(Y)$ Given $x_1 = 132$, $x_2 = 92$, in Example 26.1.1 131

27.1 Fuzzy Test $\overline{T}$ verses $\overline{CV}_2$ in Example 27.2.1($\overline{CV}_2$ left, $\overline{T}$ right) 135
27.2 Fuzzy Test $\overline{T}$ verses the $\overline{CV}_i$ in Example 27.3.1 ($\overline{CV}_1$ left, $\overline{CV}_2$ right, $\overline{T}$ center) 137

List of Tables

6.1 Values of $factor$ for Various Values of n 32

9.1 Forecast High Temperatures and Actual Values 44

19.1 SAT Scores in Example 19.2.1 96

22.1 Crisp Data for Example 22.2.1 109

23.1 Comparing the 99% Confidence Intervals in Example 23.1 . . 115

25.1 Crisp Data for Example 25.2.1 126

26.1 Comparing the 99% Confidence Intervals in Example 26.1.1 . 132